T0176010

Lectures on Ergodic Theory

Lectures on Ergodic Theory

Paul R. Halmos

Dover Publications
Garden City, New York

Bibliographical Note

This Dover edition, first published in 2017, is an unabridged republication of the work originally published in 1956 by the Chelsea Publishing Company, New York.

Library of Congress Cataloging-in-Publication Data

Names: Halmos, Paul R. (Paul Richard), 1916–2006, author.
Title: Lectures on ergodic theory / Paul R. Halmos.
Description: Garden City, New York : Dover Publications, 2017. | Series: Dover books on mathematics | Includes bibliographical references.
Identifiers: LCCN 2017025505| ISBN 9780486814896 (paperback) | ISBN 0486814890 (paperback)
Subjects: LCSH: Ergodic theory. | BISAC: MATHEMATICS / Applied.
Classification: LCC QA614 .H3 2017 | DDC 515/.48—dc23
LC record available at https://lccn.loc.gov/2017025505

Manufactured in the United States of America
81489003   2022
www.doverpublications.com

APOLOGY

The contents of this little volume are the notes on which I based my lectures in a course on ergodic theory at the University of Chicago in the Summer Quarter of 1955. Consequently this is not a formal and exhaustive monograph. I ask the reader to pretend that he is listening to a series of informal talks designed to rekindle interest in a useful but currently not fashionable part of mathematics. If the publication of these notes causes anyone to attack, and possibly even to solve, any of the fascinating open problems of ergodic theory, my purpose will have been more than accomplished.

P. R. H.

CONTENTS

INTRODUCTION

The study of topology is supposed to have started with a problem concerning the seven bridges of Koenigsberg; the study of ergodic theory originated with certain considerations in statistical mechanics. The mathematical outgrowth of the bridge problem is a theorem about odd and even vertices in graphs; the mathematical outgrowth of the gas problem is a theorem about the asymptotic behavior of measure-preserving transformations. In both cases the direct outgrowth of the original motivation is only a small part of an extensive theory. Since, nevertheless, there is some value in understanding the historical background of a theory, I shall spend the first few minutes of these lectures on a crude description of the relevant part of statistical mechanics.

Consider a mechanical system with n degrees of freedom. To be somewhat more specific, suppose that $n = 3k$ and that the system consists of k particles enclosed in a vessel in three-dimensional space. Assuming that the masses of these particles (e.g., molecules of a gas) and the forces they exert are completely known, the instantaneous state of the system can be described by giving the values of the n coordinates of position together with the corresponding n velocities. These $2n$ coordinates are not the only possible ones that can be used; for certain purposes, for instance, position and momentum are much more convenient than position and velocity.

The state of the system becomes from this point of view a point in $2n$-dimensional Euclidean space, the so-called phase space. As time goes on, the state of the system changes in accordance with the appropriate physical laws (differential equations); the entire history, past, present, and future, of the system comes to be represented by a certain trajectory in phase space. In accordance with classical, deterministic, mechanics, that entire trajectory can, in principle, be determined, once one point of it (an instantaneous state) is given. In practice we almost never have enough information for such a complete determination. The basic idea of statis-

1

tical mechanics, first proposed by Gibbs, is to abandon the deterministic study of one state (i.e., one point in phase space) in favor of a statistical study of an ensemble of states (i.e., a subset of phase space). Instead of asking " what will the state of the system be at time t? ", we should ask " what is the probability that at time t the state of the system will belong to a specified subset of phase space? ". The questions of greatest interest are the asymptotic ones : what will (probably) happen to the system as t tends to infinity?

If x_t is the point in phase space that represents the state of a particular system at time t, then, for each fixed t, the correspondence that sends x_0 onto x_t is a transformation, say T_t, so that $x_t = T_t x_0$. Since, on obvious physical grounds, $T_{s+t} = T_s T_t$, it follows that $\{T_t\}$ is a one-parameter group of transformations. (Such a group is usually called a flow.) One of the basic results of statistical mechanics is Liouville's theorem ; it asserts that if the coordinates used in the description of the phase space are appropriately chosen, then the flow in phase space leaves all volumes (i.e., $2n$-dimensional volumes) invariant. In other words, the transformations that constitute the flow are measure-preserving transformations ; the basic problem of statistical mechanics is to study the asymptotic properties of certain families of measure-preserving transformations.

Starting from a concrete, three-dimensional, physical set-up, the preceding discussion produced a rather abstract, multi-dimensional, mathematical idealization that turned out to have an important property (namely, the measure-preserving character of the flow). This property has other models that are as concrete as the ones that led to the abstraction in the first place. Consider, for example, the physical system consisting of a cocktail shaker containing ice and gin into which a few drops of vermouth have been introduced, and suppose that this system is acted upon by the flow induced by a conscientious application of a swizzle-stick. Such examples, of manifest interest, will serve to illuminate several of our considerations at the beginning.

Ergodic theory is the mathematical outgrowth of physical considerations such as I have been describing. The subject has several interesting and non-trivial theorems, and it makes contact with several other branches of mathematics (e.g., probability, topological groups, and Hilbert space).

At the same time the subject has some pathological aspects that tend to obscure the essential underlying structure. In order to emphasize theorems and examples, and to play down the pathology and the counterexamples, I have decided not to strive for the greatest generality. The plan is not to gloss over difficulties but to avoid unpleasantness that, at the beginning at any rate, is unnecessary. I shall try to state the theorems accurately and to prove them correctly, but I shall not hesitate to subject them to simplifying assumptions that might at times seem rather severe. (One virtue of this procedure is that it brings out quite clearly the unsolved problems in the field, of which, by the way, there are quite a few that are deep and interesting.) Thus if it helps to assume that a certain measure is finite (or possibly sigma-finite), or if it helps to assume that a certain topological space has a countable base, I shall not hesitate to do so.

The first simplifying assumption, the passage from the continuous to the discrete, can be made right away, and it will persist throughout these lectures. Each particular constituent of a flow, i.e., $T = T_{t_0}$ for every value of t_0, is a measure-preserving transformation. The group property of the T_t's implies that $T_{nt_0} = T^n$ for every integer n, positive, negative, or zero. (The transformation $T^0 = T_0$ is the identity.) Since it is not unreasonable to expect that the asymptotic properties of T_t are essentially the same as those of T^n, we shall restrict attention to the latter, discrete, situation, in preference to the former, continuous, one. This makes some physical sense; it says that we shall try to learn something about the asymptotic properties of the flow by taking observations at discrete, equally spaced, time intervals. It also makes mathematical sense, in that it puts the emphasis on the more primitive concept. The first objects of study are individual measure-preserving transformations; groups of such transformations will have to wait their turn.

One more word on the same subject, before I turn to the systematic presentation of the theory. The fame of one or two limit theorems has established the popular belief that they are the item of greatest interest in modern ergodic theory. I do not think that they are that. There are many algebraic and topological facts to be learned; the more intricate analytic facts acquire their proper context only when they are discussed in a general topological-algebraic structure.

EXAMPLES

The basic concept is that of a measure space, i.e., a set X together with a specified sigma-algebra of subsets of X and a measure defined on that algebra. Recall that a sigma-algebra is a class of sets closed under the formation of complements and countable unions, and that a measure is a non-negative (possibly infinite) and countably additive set function. The sets in the domain of the measure are called the measurable subsets of X. All the measurs spaces we shall consider will be assumed to be sigma-finite; we shall assume, in other words, that X is the union of countably many sets of finite measure. The purpose of this assumption is to avoid some possible pathology in connection with the Fubini theorem and the Radon-Nikodym theorem; in the presence of sigma-finiteness those theorems are smoothly applicable.

Here are some typical examples of the sort of measure spaces we shall consider. A finite-dimensional Euclidean space, with Borel measurability and Lebesgue measure. The unit interval, with the same definitions of measurability and measure. The set of all sequences $x = \{x_n\}$ of 0's and 1's, where n ranges over the set of all integers; the measurable sets are the elements of the sigma-algebra generated by sets of the form $\{x : x_n = 1\}$, and the measure is determined by the condition that its value on the intersection of k distinct generating sets is always $1/2^k$. A locally compact topological group with a countable base, with Borel measurability and Haar measure.

A measurable transformation is a mapping from a measure space into a measure space, such that the inverse image of every measurable set is measurable. A measurable transformation T from X into Y will be called invertible if there exists a measurable transformation S from Y into X such that both ST and TS are equal to the identity transformation (on their respective domains). The transformation S is uniquely determined by T; it is called the inverse of T and it is denoted by T^{-1}.

5

Most of the measurable transformations we shall consider will be measure-preserving, i.e., they will be such that the inverse image of every measurable set has the same measure as the original set. To be honest, the objects of interest are not really measure-preserving transformations, but equivalence classes of such transformations; two transformations are equivalent if they differ only on a set of measure zero. The usual password that permits the consideration of equivalence classes is " identify "; I propose to identify two measure-preserving transformations if and only if they agree almost everywhere. Observe that if a measure-preserving transformation is invertible, then its inverse is also measure-preserving. Most of the transformations that have been studied in ergodic theory are invertible measure-preserving transformations of a measure space onto itself.

A typical example of a measurable but not measure-preserving transformation on the real line is given by $Tx = 2x$; it is easy to verify that $m(T^{-1}E) = \frac{1}{2} m(E)$ for every Borel set E (where m, of course, is the measure under consideration, in this case Lebesgue measure). A closely related transformation on the unit interval is defined by $Tx = 2x \pmod 1$. To be completely explicit, I consider the half-open unit interval $[0, 1)$, and I write $Tx = 2x$ when $0 \leqq x < 1/2$, and $Tx = 2x - 1$ when $1/2 \leqq x < 1$. If $E = \left[\frac{2}{8}, \frac{5}{8}\right)$, then $T^{-1}E$ is the union of $\left[\frac{2}{16}, \frac{5}{16}\right)$ and $\left[\frac{1}{2}\left(\frac{2}{8}+1\right), \frac{1}{2}\left(\frac{5}{8}+1\right)\right)$, so that $m(T^{-1}E) = \frac{3}{16} + \frac{3}{16} = \frac{3}{8} = m(E)$. Similar considerations prove that $m(T^{-1}E) = m(E)$ whenever E is a half-open interval with dyadically rational end-points, and from there it follows easily that T is measure-preserving. Since T is not one-to-one (in fact it is everywhere two-to-one), and since it cannot be made one-to-one by any alteration on a set of measure zero, we have here an example of a measure-preserving transformation that is not invertible. An isomorphic representation of the same transformation (in an as yet undefined but pretty obvious sense) is obtained as follows. Let the measure space be the set of all complex numbers of absolute value 1, with Borel measurability, and with measure so normalized that the measure of an arc is $\frac{1}{2\pi}$ times its length; define T by $Tz = z^2$.

A simple example of an invertible measure-preserving transformation on the real line is defined by $Tx = x + 1$. More generally, in a finite-dimensional Euclidean space, let c be an arbitrary vector and define T by $Tx = x + c$. Still more generally, in a locally compact group with a left-invariant Haar measure, let c be an arbitrary element of the group and define T by $Tx = cx$. A useful special case of this last generalization is obtained by considering the circle group; in this case the geometric realization of T is the rotation through the angle arg c. An isomorphic representation of this special case can be obtained on the unit interval by selecting a number c between 0 and 1 and writing $Tx = x + c$ (mod 1). Explicitly: $Tx = x + c$ when $0 \leqq x < 1 - c$ and $Tx = x + c - 1$ when $1 - c \leqq x < 1$.

Another cluster of examples is suggested by the transformation defined on two-dimensional Euclidean space by $T(x, y) = \left(2x, \frac{1}{2}y\right)$. The inverse image of the unit square is a rectangle with base $\frac{1}{2}$ and altitude 2. Since, similarly, the inverse image of every rectangle is a rectangle of the same area, it follows that T is a measure-preserving transformation; obviously T is invertible. In an attempt to generalize this example, consider an arbitrary linear transformation T on a finite-dimensional Euclidean space. The range of T is a subspace whose inverse image is the entire space. Since a proper subspace has measure zero, it follows that in order that T be measure-preserving it is necessary that T be non-singular. If T is non-singular with determinant d, then it is well-known that $m(T^{-1}E) = m(E)/|d|$ for every Borel set E. (This well-known fact is seldom proved. A proof can be given by analytic techniques involving Jacobians; for a direct proof see Carathéodory, Vorlesungen ueber reelle Funktionen, 1927, p. 346.)

The non-singular linear transformations of a real, finite-dimensional vector space can be characterized as the continuous automorphisms of the additive vector group. This suggests the consideration of an arbitrary locally compact group with a left Haar measure and of a continuous automorphism T of that group. To find out whether or not T is measure-preserving, we must compare $m(E)$ with $n(E) = m(T^{-1}E)$. The set function n is obviously a measure; it is natural to ask if it is a left Haar measure. This means: is $m(T^{-1}(xE))$ the same as $m(T^{-1}E)$? The answer is yes, because, in fact, $T^{-1}(xE)$ is a left translate of $T^{-1}E$; indeed, an obvious computation shows that $T^{-1}(xE) = (T^{-1}x^{-1})^{-1}T^{-1}E$. It follows from the uniqueness of Haar measure that $m(T^{-1}E)$ is a constant multiple of $m(E)$.

In general this is all that we can conclude; the examples of non-singular linear transformations show that an automorphism need not be measure-preserving. If, however, the group X is compact, then $m(X)$ is finite and therefore the constant of proportionality can be evaluated by putting E equal to X; since $T^{-1}X = X$, it follows that the constant is equal to 1, and hence that T is measure-preserving.

An interesting special case is obtained by considering the group to be the torus, i.e., the Cartesian product of two circles. Concretely, the elements of the group are pairs (u, v) of complex numbers of modulus one; the group operation is coordinatewise multiplication. It is easy to show that the most general continuous automorphism is given by a two-rowed unimodular matrix, i.e., by a matrix with integer entries and determinant ± 1; if $\begin{pmatrix} a & b \\ c & d \end{pmatrix}$ is such a matrix, the corresponding automorphism T is defined by $T(u, v) = (u^a v^b, u^c v^d)$.

Let X be the space of sequences $x = \{x_n\}$, $n = 0, \pm 1, \pm 2, \cdots$, described before; let T be the transformation induced by a unit shift on the indices, i.e., $Tx = y = \{y_n\}$, where $y_n = x_{n+1}$. This transformation is measure-preserving and invertible. If X is the unilateral sequence space, i.e., the elements of X are sequences $\{x_n\}$ with $n = 0, 1, 2, \cdots$, the same equation defines a measure-preserving but non-invertible (two-to-one) transformation.

There is a simple mapping S from the unilateral sequence space to the unit interval; S sends the sequence $\{x_n\}$ of 0's and 1's onto the number whose binary expansion is $.x_1 x_2 \cdots$. The transformation S is measure-preserving and essentially one-to-one. It is not quite one-to-one, because a dyadically rational number has two possible expansions. The set of sequences whose image is dyadically rational has the same cardinal number as the set of dyadically rational numbers; both sets are countably infinite. If we suitably redefine S on these exceptional sequences, the result is an invertible measure-preserving transformation from the sequence space onto the unit interval. The existence of such a transformation shows that the measure-theoretic structures of the two spaces are isomorphic. The isomorphism (i.e., the transformation S) carries the unilateral shift T onto an invertible measure-preserving transformation T' on the interval; T' is defined by $T' = STS^{-1}$. An examination of the definitions of S and T shows that T' is an old friend: $T'x = 2x \pmod 1$ almost everywhere.

There is a natural correspondence between the bilateral sequence space and the Cartesian product of the unilateral sequence space with itself; the correspondence sends $\{\cdots, x_{-2}, x_{-1}, x_0, x_1, x_2, \cdots\}$ onto

$$(\{x_0, x_1, x_2, \cdots\}, \{x_{-1}, x_{-2}, \cdots\}).$$

This correspondence is easily seen to be an invertible measure-preserving transformation, and, therefore, a measure-theoretic isomorphism. If we denote this isomorphism by P and if we denote by Q the Cartesian product of S with itself (so that $Q(x, y) = (Sx, Sy)$ whenever x and y are unilateral sequences), then the composite transformation QP is an isomorphism from the bilateral sequence space onto the unit square. This isomorphism carries the bilateral shift onto an invertible measure-preserving transformation T'' on the square. An examination of the definitions shows that T'' is a close relative of an old friend; it is given by $T''(x, y) = \left(2x, \frac{1}{2}y\right)$ when $0 \leqq x < \frac{1}{2}$ and by $T''(x, y) = \left(2x, \frac{1}{2}(y+1)\right)$ when $\frac{1}{2} \leqq x < 1$. (These equations, valid almost everywhere, must of course be taken modulo 1.) The transformation T'' can be described geometrically, as follows. Transform the unit square by the linear transformation that sends (x, y) onto $\left(2x, \frac{1}{2}y\right)$, getting a rectangle whose bottom edge is $[0, 2)$ and whose left edge is $\left[0, \frac{1}{2}\right)$; cut off the right half of this rectangle (with bottom edge $[1, 2)$) and move it, by translation, to the top half of the unit square. Since these actions are faintly reminiscent of what happens in kneading dough, the transformation T'' is sometimes called the baker's transformation.

RECURRENCE

In order to discuss the asymptotic properties of a measure-preserving transformation T, i.e., the properties of the sequence $\{T^n\}$, the powers of T must make sense; for this reason we shall, throughout the sequel, restrict attention to transformations from a set X into itself. The earliest and simplest asymptotic questions were raised by Poincaré (Calcul des probabilités, 1912); they concern recurrence. If T is a measurable transformation on X and E is a measurable subset of X, a point x of E is called recurrent (with respect to E and T) if $T^n x \in E$ for at least one positive integer n. Our first result is typical of the subject.

RECURRENCE THEOREM. *If T is a measure-preserving (but not necessarily invertible) transformation on a space of finite measure, and if E is a measurable set, then almost every point of E is recurrent.*

PROOF. If not, then the set F of those points of E that never return to E is a set of positive measure. The set F is measurable since

$$F = E \cap T^{-1}(X - E) \cap T^{-2}(X - E) \cap \cdots.$$

If $x \in F$, then none of the points Tx, $T^2 x$, $T^3 x, \cdots$, belongs to F, or, in other words, F is disjoint from $T^{-n}F$ for all positive n. It follows that the sets F, $T^{-1}F$, $T^{-2}F, \cdots$, are pairwise disjoint, since

$$T^{-n}F \cap T^{-(k+n)}F = T^{-n}(F \cap T^{-k}F).$$

Since T is measure-preserving and the space has finite msasure, this is a contradiction.

The recurrence theorem implies a stronger version of itself. Not only is it true that for almost every x in E at least one term of the sequence Tx, $T^2 x, \cdots$ belongs to E; in fact, for almost every x in E, there are infinitely many values of n such that $T^n x \in E$. The idea of the proof is to apply the recurrence theorem to each power of T. Precisely speaking, if F_n is the set of those points of E that never return to E under the action of T^n, then, by the recurrence theorem, $m(F_n) = 0$. If $x \in E - (F_1 \cup F_2 \cup \cdots)$, then $T^n x \in E$ for some positive n, since $x \in E - F_1$. Similarly, since $x \in E - F_n$, it follows that $T^{kn} x \in E$ for some positive k. The strengthened version of the recurrence theorem follows by an inductive repetition of this twice-repeated argument.

In the proof of the original recurrence theorem, the measure-preserving character of the transformation and the finiteness of the measure were used in a very weak sense only. All that was essential was the non-existence of a set F of positive measure such that the sets F, $T^{-1}F$, $T^{-2}F$, $\cdots$ are pairwise disjoint. Motivated by this remark, we introduce a new concept : a measurable transformation T is called dissipative if there exists a measurable set F of positive measure such that the sets F, $T^{-1}F$, $T^{-2}F, \cdots$ are pairwise disjoint; in the contrary case T is called conservative. It is clear that the weak recurrence theorem is valid for every conservative transformation.

The proof of the strong recurrence theorem depends on the applicability of the weak theorem to every power of T. It is clear that if T is dissipative, then so is every power of T. This is not good enough. If we knew that every power of a conservative transformation is itself conservative, we could assert the conclusion of the strong recurrence theorem for every conservative transformation.

The question is partially clarified by still another definition : we shall say that T is compressible if there exists a measurable set E such that $E \subset T^{-1}E$ and such that $m(T^{-1}E - E) > 0$. In the contrary case T is called incompressible. The virtue of compressibility is that it is slightly easier to work with, and at the same time equivalent to, dissipation. Indeed, if $E \subset T^{-1}E$ with $m(T^{-1}E - E) > 0$, write $F = T^{-1}E - E$; it follows that $\{T^{-n}F\}$ is a disjoint sequence. If, on the other hand, F is a set of positive measure such that $\{T^{-n}F\}$ is disjoint, then write $E = X - (F \cup T^{-1}F \cup T^{-2}F \cup \cdots)$; it follows that $E \subset T^{-1}E$ and $m(T^{-1}E - E) > 0$.

I proved once (Annals, 1947, p. 738) that if T is one-to-one and if T^{-1} is measurable, then the incompressibility of T implies that of every power of T. The proof is combinatorial and somewhat intricate. It should not be difficult to find out whether the result remains true for transformations that are not one-to-one; it is likely that a minor modification of the proof for the one-to-one case can be made to apply.

In connection with compressibility, it is worth mentioning that a one-to-one measurable transformation with a measurable inverse on a space of finite measure always has an essentially uniquely determined incompressible part. More precisely: there exists a measurable invariant set Y (i.e., $T^{-1}Y = Y$) such that T is incompressible on Y, and there exists a measurable dissipative set F such that $X - Y$ is the union of the sets T^nF, $n=0$, $\pm 1, \pm 2, \cdots$. (Idea of the proof: let d be the supremum of $m(D)$ for all dissipative sets D, find a sequence $\{D_n\}$ of dissipative sets such that $m(D_n)$ tends to d, let Y be the complement of the least invariant set including all the D_n's.) The generalization of this result to transformations that are not necessarily one-to-one is likely to be delicate. Consider, for an example, the set of non-negative integers in the role of X; write $Tx = x - 1$ when $n > 1$ and $T1 = 3$.

The preceding discussion hinted at two possibilities, both of which are realizable. First, a measure-preserving transformation on a space of infinite measure need not be conservative, not even if it is invertible. Example: $Tx = x + 1$ on the discrete space of the integers. The recurrence theorem is false for this T. Second, a one-to-one measure-preserving transformation need not be invertible. Example: let X be the set of integers; call a subset of X measurable if and only if its intersection with the set of non-negative integers is invariant under all the transpositions that interchange $2n$ with $2n+1$, $n=0, 1, 2, \cdots$; write $Tx = x + 2$. This example shows also that the union of two compressible sets need not be compressible; consider the non-negative integers with the even negative integers for one set and the non-negative integers with the odd negative integers for another.

The conclusion of the recurrence theorem can be formulated in terms of the characteristic function, say f, of the set E, as follows: for almost every x in E the series $\sum f(T^nx)$ diverges. This conclusion can be generalized: if f is an arbitrary non-negative measurable function, then for almost every x in the set $\{x : f(x) > 0\}$ the series $\sum f(T^nx)$ diverges. The proof is easy. Consider, for every positive integer k, the set E_k where $f(x) > \frac{1}{k}$. The recurrence theorem implies that for almost every x in E_k the point T^nx will belong to E_k infinitely often; the desired result follows by forming the union of the E_k's.

MEAN CONVERGENCE

The recurrence theorem says that under appropriate conditions on a transformation T almost every point of each measurable set E returns to E infinitely often. It is natural to ask: exactly how long a time do the images of such recurrent points spend in E? The precise formulation of the problem runs as follows: given a point x (for present purposes it does not matter whether x is in E or not), and given a positive integer n, form the ratio of the number of these points that belong to E to the total number (i.e., to n), and evaluate the limit of these ratios as n tends to infinity. It is, of course, not at all obvious in what sense, if any, that limit exists.

If f is the characteristic function of E, then the ratio just discussed is $\frac{1}{n}\sum_{j=0}^{n-1}f(T^jx)$. The problem of the mean time of sojourn is therefore the problem of Cesàro convergence for the sequence $\{f(T^nx)\}$. The first significant step in this direction was made soon after it was recognized that it is neither necessary nor desirable to restrict attention to characteristic functions.

If f is an arbitrary function on X, another function g on X is defined by writing $g(x)=f(Tx)$. If we write $g=Uf$, then U is a mapping that operates on functions. The mapping U has some important properties. The most obvious property of U is its linearity: if, for instance, f and g are complex-valued functions on X and if a and b are complex scalars, then $U(af+bg)=aUf+bUg$. If T is measure-preserving, then U has the more important property that it sends L_1 into itself, and, in fact, is an isometry on L_1. What this means, of course, is that if $f\in L_1$, then $Uf\in L_1$ and $\|f\|_1=\|Uf\|_1$, where, in general, $\|f\|_p$ denotes L_p norm. The proof is easy. If f is the characteristic function of a set E of finite measure, then Uf is the characteristic function of $T^{-1}E$; the desired result in this case follows from the fact that $\|f\|_1=m(E)$. From this and from the linearity

13

of U it follows that U is norm-preserving on finite linear combinations of such characteristic functions, i.e., on simple functions. If f is a non-negative function, then f is the pointwise limit of an increasing sequence $\{f_n\}$ of non-negative simple functions. Since $\{Uf_n\}$ is also an increasieg sequence of non-negative functions, it follows from the theorem on integration of monotone sequences that we have $\lim \|Uf_n\|_1 = \|Uf\|_1$ as well as $\lim \|f_n\|_1 = \|f\|_1$; this proves the result for non-negative functions. The general case now follows from the fact that the norm of every f in L_1 is the same as the norm of $|f|$. Observe that in none of these considerations is it necessary to assume that the underlying space has finite measure.

The fact that U is an isometry on L_1 implies immediately that U is an isometry on L_2; all that is needed is the observation that the L_2 norm of f is the square root of the L_1 norm of f^2. If T is an invertible measure-preserving transformation, then U is an invertible isometry; the inverse of U is the operator V defined by $Vf(x)=f(T^{-1}x)$. An invertible isometry on a Hilbert space is a unitary operator, so that what we have proved is that the functional operator induced on L_2 by an invertible measure-preserving transformation is unitary. This fact was first pointed out by Koopman (Proc. N.A.S., 1931, p. 315).

A basic asymptotic problem of ergodic theory reduces thus to studying the limiting behavior of the averages $\frac{1}{n}\sum_{j=0}^{n-1}U^j$, where U is an isometry on a Hilbert space. In Hilbert space terms, however, the natural question is not that of the pointwise convergence of $\frac{1}{n}\sum_{j=0}^{n-1}f(T^jx)$ but rather its convergence in the mean (of order two). The assertion that mean convergence always does take place is the first result of modern ergodic theory. This result (for unitary operators) was first proved by von Neumann; cf. the historical remarks of G. D. Birkhoff and Koopman (Proc. N.A.S., 1932, p. 281). The theorem of von Neumann is called the mean ergodic theorem, in distinction from Birkhoff's subsequent result, the so-called individual ergodic theorem.

The mean ergodic theorem is an amusing and simple piece of classical analysis in case the underlying Hilbert space is one-dimensional. In that case an isometry is determined by a complex number u of absolute value 1; the problem is the convergence of $\frac{1}{n}\sum_{j=0}^{n-1} u^j$. If $u=1$, each average is equal to 1. If $u\neq 1$, the n-th average is $(1-u^n)/n(1-u)$, and therefore the absolute value of the n-th average is dominated by $\frac{2}{n(1-u)}$; it follows that the averages tend to 0.

More generally, in the finite-dimensional case, every isometry is given by a unitary matrix, which, without any loss of generality, may be assumed to be a diagonal matrix. Since the diagonal entries of such a matrix U are complex numbers of absolute value 1, it follows that the averages tend to a diagonal matrix; the diagonal entries of the limit are 1's and 0's. The limit matrix, say P, is therefore a projection; it is, in fact, the projection on the space of all vectors f such that $Uf=f$.

A simple and ingenious proof of the mean ergodic theorem in the general case was devised by F. Riesz. In the course of that proof I shall need to make use of a simple fact about isometries on a complex Hilbert space. Before stating that fact, I want to establish the Hilbert-space-theoretic notation to be used throughout. The norm of a vector f in a Hilbert space will always be denoted by $\|f\|$; if the Hilbert space is L_2, the subscript in $\|f\|_2$ will generally be omitted. The inner product of f and g will be denoted by (f,g); in the measure-theoretic case $(f,g) = \int f(x)\overline{g(x)}\,dx$. (Observe, incidentally, the notation for integrals. Whenever there is no doubt as to what measure is being used, I shall write $\int f(x)\,dx$ instead of $\int f(x)\,dm(x)$. When the domain of integration is not explicitly indicated, it is always understood to be the entire space.) The adjoint of an operator U will be denoted by U^*; it is characterized by the equation $(Uf, g) = (f, U^*g)$, valid for all f and g. The identity operator on a Hilbert space will be denoted by 1.

The auxiliary fact about isometries is this: if U is an isometry, then a necessary and sufficient condition that $Uf=f$ is that $U^*f=f$. Indeed, if $Uf=f$, then, applying U^* to both sides, I conclude that $f=U^*f$. (An operator U is an isometry, i.e., $\|Uf\| = \|f\|$ for all f, if and only if $U^*U = 1$. The proof is the same in general as it is in the finite-dimensional case.   Caution : in the finite-dimensional case $U^*U = 1$ implies $UU^* = 1$; in the general case this is not true.   In other words, an isometry on an infinite-dimensional Hilbert space need not be unitary.)   If, conversely, $U^*f = f$, then I show that $\|Uf - f\| = 0$.   Indeed :

$$\|Uf - f\|^2 = (Uf - f, Uf - f) = \|Uf\|^2 - (f, Uf) - (Uf, f) + \|f\|^2.$$

Since $(f, Uf) = (U^*f, f) = \|f\|^2$, and, similarly, $(Uf, f) = (f, U^*f) = \|f\|^2$, the proof is complete.

We are now ready for the slightly generalized version of von Neumann's theorem.

MEAN ERGODIC THEOREM.   *If U is an isometry on a complex Hilbert space and if P is the projection on the space of all vectors invariant under U, then $\dfrac{1}{n}\displaystyle\sum_{j=0}^{n-1} U^j f$ converges to Pf for every f in the space.*

PROOF.   If $Uf = f$, then $\dfrac{1}{n}\displaystyle\sum_{j=0}^{n-1} U^j f$ obviously converges to f.   If $f = g - Ug$ for some g, then $\displaystyle\sum_{j=0}^{n-1} U^j f$ is a telescoping sum equal to $g - U^n g$; it follows that

$$\left\| \frac{1}{n}\sum_{j=0}^{n-1} U^j f \right\| \leq \frac{2}{n}\|g\|,$$

and hence that the averages tend to 0.   The rest of the proof shows that in a certain sense every f is an easy combination of f's for which $Uf = f$ and of f's of the form $g - Ug$.

The set of f's of the form $g - Ug$ is a linear manifold that is not necessarily closed.   The uniform boundedness of the averages $A_n = \dfrac{1}{n}\displaystyle\sum_{j=0}^{n-1} U^j$ implies that $A_n f$ tends to 0 for every f in the closure of that manifold. More generally : if f_k tends to f and if $A_n f_k$ tends to 0 for each k, then $A_n f$ tends to 0.   Proof : note that

$$\|A_n f\| \leq \|A_n(f - f_k)\| + \|A_n f_k\|,$$

select k so as to make $\|f - f_k\|$ small, conclude that $\|A_n(f - f_k)\|$ is just as small (no matter what n may be), and then select n so as to make $\|A_n f_k\|$ small.

The orthogonal complement of the set of f's of the form $g - Ug$ is the same as the orthogonal complement of the closure of that set. If h belongs to that orthogonal complement, so that $(h, g - Ug) = 0$ for all g, then $(h, g) - (U^*h, g) = 0$, or $(h - U^*h, g) = 0$ for all g. It follows that $h = U^*h$ and hence that $Uh = h$. The steps of the argument are reversible, so that if $Uh = h$, then $(h, Ug - g) = 0$ for all g. Conclusion : the orthogonal complement of the set of f's of the form $g - Ug$ is the set of invariant f's. It follows that every f can be written as a sum $f_1 + f_2$, where $Uf_1 = f_1$ and where f_2 is in the closure of the $(Ug - g)$'s, and this completes the proof of the mean ergodic theorem.

The techniques and results here described have been generalized to large classes of operators and groups of operators on large classes of abstract vector spaces. I do not propose to enter into a discussion of these generalizations; I shall, instead, turn to the more delicate measure-theoretic problems.

POINTWISE CONVERGENCE

We turn now to Birkhoff's ergodic theorem. The proof is (again) due to F. Riesz (Commentarii, 1945, p. 221). It begins with an amusing piece of combinatorics.

Suppose that $a_1, \cdots, a_n$ is a finite sequence of real numbers and that m is a positive integer, $m \leq n$. A term a_k of the sequence will be called an m-leader if there exists a positive integer p such that $1 \leq p \leq m$ and such that $a_k + \cdots + a_{k+p-1} \geq 0$. Thus, for instance, the 1-leaders are the non-negative terms of the sequence; observe, however, that if $m > 1$, an m-leader need not be non-negative.

LEMMA. *The sum of the m-leaders is non-negative.*

PROOF. If there are no m-leaders, the assertion is true. Otherwise, let a_k be the first m-leader and let $a_k + \cdots + a_{k+p-1}$ be the shortest non-negative sum that it leads ($p \leq m$). I assert that every a_h in this sum is itself an m-leader, and, in fact, that $a_h + \cdots + a_{k+p-1} \geq 0$. Indeed, if not, then $a_k + \cdots + a_{h-1} > 0$, contradicting the original choice of p. Proceed now inductively with the sequence $a_{k+p}, \cdots, a_n$; the sum of the shortest non-negative sums so obtained is exactly the sum of the m-leaders.

INDIVIDUAL ERGODIC THEOREM. *If T is a measure-preserving (but not necessarily invertible) transformation on a space X (with possibly infinite measure) and if $f \in L_1$, then $\frac{1}{n} \sum_{j=0}^{n-1} f(T^j x)$ converges almost everywhere. The limit function f^* is integrable and invariant (i.e., $f^*(Tx) = f^*(x)$ almost everywhere). If $m(X) < \infty$, then $\int f^*(x)\, dx = \int f(x)\, dx$.*

PROOF. There is obviously no loss of generality in assuming that f is real-valued. For convenience in writing, I shall use $f_j(x)$ as an abbreviation for $f(T^j x)$. The proof begins by establishing an auxiliary statement that has acquired a dignified name of its own.

MAXIMAL ERGODIC THEOREM. *If E is the set of those points x for which*

18

at least one of the sums $f_0(x) + \cdots + f_{n-1}(x)$ *is non-negative, then* $\int_E f(x) dx \geqq 0$.

Let E_m be the set of those points x for which at least one of the sums $f_0(x) + \cdots + f_p(x)$ is non-negative with $p \leqq m$. Clearly the sequence of E's is increasing and its union is exactly E; it is sufficient therefore to prove that $\int_{E_m} f(x) dx \geqq 0$ for each m.

Let n be an arbitrary positive integer and consider for each point x the m-leaders of the sequence $f_0(x), \cdots, f_{n+m-1}(x)$; let $s(x)$ be their sum. Let D_k be the set of those points x for which $f_k(x)$ is an m-leader of the sequence $f_0(x), \cdots, f_{n+m-1}(x)$ and let g_k be its characteristic function. Since D_k is measurable and $s = \sum_{k=0}^{n+m-1} f_k g_k$, it follows that s is measurable and integrable. The lemma implies therefore that

(*) $$\sum_{k=0}^{n+m-1} \int_{D_k} f_k(x) dx \geqq 0.$$

Observe next that if $k = 1, \cdots, n-1$, then the following conditions on the point x are mutually equivalent: (i) $Tx \in D_{k-1}$, (ii) $f_{k-1}(Tx) + \cdots + f_{k-1+p-1}(Tx) \geqq 0$ for some $p \leqq m$, (iii) $f_k(x) + \cdots + f_{k+p-1}(x) \geqq 0$ for some $p \leqq m$, (iv) $x \in D_k$. In other words, $D_k = T^{-1}D_{k-1}$, or $D_k = T^{-k}D_0$ for $k = 1, \cdots, n-1$.

Consequently

$$\int_{D_k} f_k(x) dx = \int_{T^{-k}D_0} f(T^k x) dx = \int_{D_0} f(x) dx$$

(by the change of variables that replaces $T^k x$ by x), so that the first n terms of the sum(*) are all equal to the first one. Since clearly $D_0 = E_m$, (*) implies that

$$n \int_{E_m} f(x) dx + m \int |f(x)| dx \geqq 0.$$

(The last m terms of (*) were replaced by obvious majorants.) The proof of the maximal ergodic theorem is completed by dividing by n and then allowing n to tend to infinity.

Now to prove the ergodic theorem itself. If a and b are real numbers with $a < b$, let $Y = Y(a, b)$ be the set of those points of x for which

$$\liminf \frac{1}{n} \sum_{j=0}^{n-1} f_j(x) < a < b < \limsup \frac{1}{n} \sum_{j=0}^{n-1} f_j(x).$$

Clearly Y is measurable and invariant under T (i.e., $Y = T^{-1}Y$); it will now be proved that $m(Y)$ is finite, and then that $m(Y) = 0$.

I may assume that $b > 0$; if this is not the case, then $a < 0$ and the argument may be carried out with $-f$ and $-a$ in place of f and b respectively. Let C be an arbitrary subset of Y, measurable and of finite measure. Let g be the characteristic function of C and apply the maximal ergodic theorem to $f - bg$ in place of f. If F denotes the set that plays for this function the same role as E played for f, the conclusion becomes $\int_F (f(x) - bg(x)) \, dx \geqq 0$. If $x \in Y$, so that $b < \limsup \frac{1}{n} \sum_{j=0}^{n-1} f_j(x)$, then at least one of the averages $\frac{1}{n} \sum_{j=0}^{n-1} f_j(x)$ must be greater than b (in fact many of them must be such); it follows that at least one of the sums $\sum_{j=0}^{n-1} (f(T^j x) - bg(T^j x))$ is non-negative. In other words, $Y \subset F$. Using this and the maximal ergodic theorem, I conclude that $\int_F f(x) \, dx \geqq \int_F bg(x) \, dx$ and hence that $\int |f(x)| \, dx \geqq bm(C)$. I have proved thus that if a measurable subset of Y has finite measure, then that measure is dominated by $\frac{1}{b} \int |f(x)| \, dx$; it follows (by sigma-finiteness) that $m(Y) < \infty$. Since Y is invariant under T, the space X may be replaced by Y and the maximal ergodic theorem may then be applied to the (integrable!) function $f - b$; since the set E of the maximal ergodic theorem coincides in this case with Y, it follows that $\int_Y (f(x) - b) \, dx \geqq 0$. A similar application of the maximal ergodic theorem to $a - f$ yields $\int_Y (a - f(x)) \, dx \geqq 0$. The sum of the last two inequalities is $\int_Y (a - b) \, dx \geqq 0$; the hypothesis $a < b$ implies therefore that $m(Y) = 0$.

Applying the result so obtained to all pairs (a, b) of rational numbers $(a < b)$, it follows that the averages do indeed converge to a limit almost everywhere. Since, moreover,

$$\int \left| \frac{1}{n} \sum_{j=0}^{n-1} f_j(x) \right| \, dx \leqq \frac{1}{n} \int \sum_{j=0}^{n-1} |f_j(x)| \, dx = \int |f(x)| \, dx < \infty,$$

the limit function f^* is integrable (Fatou's lemma), and therefore finite almost everywhere. The fact that f^* is invariant is a trivial consequence of the elementary properties of Cesàro convergence.

It remains only to prove that if $m(X)$ is finite, then f and f^* have the same integral. This is another application of the maximal ergodic theorem. If $f^*(x) \geq a$ everywhere, then at least one of the sums $\sum_{j=0}^{n-1}(f_j(x) - a + \varepsilon)$ must be non-negative for each ε; it follows that $\int f(x)dx \geq (a - \varepsilon)m(X)$ for each ε, and hence that $\int f(x)dx \geq am(X)$. Similarly, of course, if $f^*(x) \leq b$ everywhere, then $\int f(x)dx \leq bm(X)$. Write $X(k, n)$ for the set of x's where $k/2^n \leq f^*(x) \leq (k+1)/2^n$ and apply these inequalities to the transformation T restricted to the (invariant) set $X(k, n)$. It follows that

$$\frac{k}{2^n} m(X(k, n)) \leq \int_{X(k, n)} f(x)\,dx \leq \frac{k+1}{2^n} m(X(k, n)).$$

These latter inequalities are valid for f^* also. It follows that

$$-\frac{1}{2^n} m(X(k, n)) \leq \int_{X(k, n)} f(x)\,dx - \int_{X(k, n)} f^*(x)\,dx \leq \frac{1}{2^n} m(X(k, n));$$

summing over k, I obtain

$$\left| \int f(x)\,dx - \int f^*(x)\,dx \right| \leq \frac{1}{2^n} m(X).$$

Since n is arbitrary, the proof of the ergodic theorem is complete.

COMMENTS ON THE ERGODIC THEOREM

The analytic trickery is over; it is time to harvest the corollaries. The first corollary is that on a space of finite measure we have convergence in the mean (of order one), as well as almost everywhere convergence. If, in other words, T is a measure-preserving transformation on X, with $m(X) < \infty$, and if $f \in L_1$, then $\int \left| \frac{1}{n} \sum_{j=0}^{n-1} f(T^j x) - f^*(x) \right| dx$ tends to 0, where, of course, $f^*(x) = \lim \frac{1}{n} \sum_{j=0}^{n-1} f(T^j x)$. If f is bounded, then the averages all have the same bound and the assertion follows from the Lebesgue bounded convergence theorem. If f is not bounded, the assertion follows from an approximation argument. If g is a bounded function, then

$$\left\| \frac{1}{n} \sum_{j=0}^{n-1} f_j - f^* \right\|_1 \leq \left\| \frac{1}{n} \sum_{j=0}^{n-1} (f_j - g_j) \right\|_1 + \left\| \frac{1}{n} \sum_{j=0}^{n-1} g_j - g^* \right\|_1 + \| g^* - f^* \|_1.$$

The first term on the majorant side is dominated by $\|f - g\|_1$, and the third term is equal to $\|f - g\|_1$. Consequently if g is selected so as to make $\|f - g\|_1$ small, and then n is selected so as to make the middle term small, it follows that the minorant term is small, as asserted.

A popular aspect of analytic ergodic theory is the investigation of inter-deducibility techniques such as the one just described. (Does the L_2 theorem follow from the almost everywhere theorem? ; does the almost everywhere theorem follow from the L_1 theorem? ; etc.). Similar techniques are used to generalize the ergodic theorems to transformations that are not necessarily measure-preserving. Since these elaborations are yet to produce any worth while applications, I do not enter into them here.

The generalization to a continuous parameter is deserving of mention; it has one new aspect that must be kept in mind. The obvious thing is to consider a one-parameter semigroup $\{T_t\}$ of measure-preserving

22

transformations, where t is a non-negative real number and where $T_{s+t} = T_s T_t$. The sums over powers of a transformation that occur in the discrete ergodic theorems become integrals in the continuous case; the ergodic theorem asserts the convergence (as N becomes infinite) of $\frac{1}{N}\int_0^N f(T_t x)dt$, where f is an arbitrary element of L_1. In order for the integral that occurs here to make sense, some assumption has to be made on the way T_t depends on t. The natural assumption turns out to be that $T_t x$ should be a measurable function of its two arguments (i.e., t and x), where measurability on the positive real axis is interpreted in the sense of Borel. Under this assumption the continuous ergodic theorem is meaningful and true. The proof is a straightforward imitation of the proof in the discrete case; alternatively the continuous case can be reduced to the discrete case. The essential trick in the second method is to apply the discrete ergodic theorem to the transformation T_1 and to the function F defined by $F(x)=\int_0^1 f(T_t x)dt$.

The assertion that $\int f(x)dx = \int f^*(x)dx$ was proved only for a space of finite measure. The assumption of finiteness cannot be omitted here; for spaces of infinite measure the conclusion is false. If, for instance, X is the real line and T is the translation defined by $Tx = x+1$, and if f is the characteristic function of the half-open unit interval, then $\int f(x)dx = 1$, whereas $f^*(x)=0$ for all x.

Another natural question concerns the converse of the ergodic theorem, in the following sense: if T is a measure-preserving transformation and if f is a measurable function such that $\frac{1}{n}\sum_{j=0}^{n-1} f(T^j x)$ converges to a finite limit $f^*(x)$ almost everywhere, does it follow that f is integrable? The answer is no in general; a little later I shall give an example of an interesting special case where the answer is yes. For a negative example, consider again the translation by a unit amount on the real line; if f is an arbitrary function that vanishes outside the unit interval (in particular f may be non-negative and non-integrable), then f^* is identically zero. The most extreme counterexample is given by the identity transformation; for it the conclusion of the individual ergodic theorem always holds.

I cannot resist the temptation of concluding these comments with an alternative " proof " of the ergodic theorem.   If f is a complex-valued function on the non-negative integers, write $\int f(n)dn = \lim \frac{1}{n}\sum_{j=0}^{n-1} f(j)$ whenever the limit exists, and call such functions integrable.   If T is a measure-preserving transformation on a space X and if f is an integrable function on X, then

$$\iint |f(T^n x)|\, dndx = \iint |f(T^n x)|\, dxdn = \iint |f(x)|\, dxdn = \int |f(x)|dx < \infty.$$

Hence, by " Fubini's theorem " (!), $f(T^n x)$ is an integrable function of its two arguments, and therefore, for almost every fixed x, it is an integrable function of n.   Can any of this nonsense be made meaningful?

ERGODICITY

If T is a measure-preserving transformation on X and if X is the union of two disjoint measurable sets E and F of positive measure, each of which is invariant under T, then the study of any property of T on X reduces to the separate studies of the corresponding properties of T on E and T on F. In such a situation T may be called decomposable. The most significant transformations are the indecomposable ones – they are usually called metrically transitive or ergodic. Ergodicity is one of the precise formulations (but not the only one) of the natural requirement that a transformation do a good job of stirring up the points of the space it acts on.

In order to give some examples of ergodic transformations, it is convenient to reformulate the definition of ergodicity. The first reformulation is obvious : T is ergodic if and only if it has only trivial invariant sets, i.e., if and only if either $m(E) = 0$ or $m(X - E) = 0$ whenever E is a measurable set invariant under T. (Recall the definition of invariance. A set E is invariant under T if and only if $T^{-1}E = E$; this means that x belongs to E if and only if Tx belongs to E. A function f is invariant under T if and only if $f(Tx) = f(x)$ for all x. Clearly E is invariant if and only if its characteristic function is invariant. I shall generally use " invariant " to mean " invariant almost everywhere ", so that for functions, for example, invariance will mean that $f(Tx) = f(x)$ for almost all x.) The useful reformulation of ergodicity is this : T is ergodic if and only if every measurable invariant function is a constant. One way the proof is trivial : if there are no non-constant invariant functions, then there are no non-trivial invariant sets. (Consider characteristic functions.) Suppose, conversely, that T is ergodic; it is to be proved that if f is measurable and invariant, then f is constant. If $X(k, n)$ is the set of x's where $k/2^n \leqq f(x) < (k+1)/2^n$, then the invariance of f implies the invariance of $X(k, n)$. The ergodicity of T implies now that for each fixed n all but one of the sets $X(k, n)$ has

25

measure zero. The desired conclusion follows by forming the intersection (over all n) of the "large" ones among the sets $X(k, n)$. In case $m(X)$ $<\infty$, it is also true that T is ergodic if and only if every invariant function in L_1 (or, for that matter, in L_2) is constant; the point is that, in that case, the characteristic function of every measurable set is integrable.

The translation T defined by $Tx = x + 1$ on the space of integers is ergodic; the translation T defined by $Tx = x + 2$ is not. (The set of even integers is invariant.) The translation T defined by $Tx = x + 1$ on the real line is not ergodic; an example of a non-trivial invariant set is the union (over all n) of the set of all x's for which $n < x < n + \frac{1}{2}$.

If X is the circle group (i.e., the set of all complex numbers of absolute value 1), if $c \in X$, and if T is defined by $Tx = cx$, then T is ergodic for some values of c and not ergodic for others. If c is a root of unity, i.e., $c^n = 1$ for some positive integer n, then T is not ergodic; indeed if $f(x) = x^n$, then f is a non-constant measurable invariant function. If c is not a root of unity, then T is ergodic. One way to prove this is to observe that the functions f_n defined by $f_n(x) = x^n$, $n = 0, \pm 1, \pm 2, \cdots$, form a complete orthonormal set in L_2. It follows that if f is in L_2, then $f = \sum_n a_n f_n$, where the series converges in the mean of order two. Define the functional operator U as before $(Uf(x) = f(Tx))$; since $Uf_n = c^n f_n$, it follows that $Uf = \sum_n a_n c^n f_n$. If f is invariant, then $a_n = a_n c^n$ for all n, and hence $a_n = 0$ whenever $n \neq 0$. This implies that every invariant function in L_2 is a constant and hence that T is ergodic.

If, more generally, X is a compact abelian group with a countable base, and if $Tx = cx$ for some c in X, then a necessary and sufficient condition that T be ergodic is that the sequence $\{c^n\}$ of powers of c, $n = 0, \pm 1, \pm 2, \cdots$, be everywhere dense in X. The proof of this fact is an interesting digression from the main body of the theory; it runs as follows.

LEMMA. *If the measure space X is a topological space with a countable base, such that every non-empty open set has positive measure, and if T is an ergodic measure-preserving transformation on X, then for almost every x in X the orbit of x (i.e., the sequence $\{T^n x\}$) is everywhere dense.*

PROOF. The orbit of x is not dense if and only if there exists a non-empty basic open set G such that x belongs to the intersection of all $X - T^n G$. Since this intersection is an invariant set disjoint from G, and since $m(G) > 0$, it follows that it has measure zero. If x does not belong to any of this countable class of sets of measure zero (one for each basic open set), then x has a dense orbit.

The density condition that this lemma asserts to be necessary for ergodicity is not sufficient. To get a counterexample, let T be a transformation on, say, the interval $[0, 2)$, such that T leaves both $[0, 1)$ and $[1, 2)$ invariant, and such that the restrictions of T to these subintervals are ergodic. The desired counterexample can be obtained by defining a topology on $[0, 2)$ (different from the usual topology, of course) so that both the subintervals $[0, 1)$ and $[1, 2)$ are dense in $[0, 2)$, and so that the resulting topological measure space satisfies the conditions of the lemma. For this purpose, consider a square in the plane and find two disjoint countable classes of half-open intervals dense in it. The unions of these two classes may be put into one-to-one correspondence with the intervals $[0, 1)$ and $[1, 2)$ in an obvious manner; let the topology on $[0, 2)$ be defined to be the one inherited from the square, via these correspondences. Alternatively, define measure in the square via these correspondences, assigning to the set of irrelevant points of the square the measure zero, and define a transformation in the square as the identity on irrelevant points and the transferred version of T elsewhere.

Suppose now that T is a rotation (i.e., $Tx = cx$ on a compact abelian group X with a countable base). If T is ergodic, then, by the lemma, there exists at least one point, say x_0, whose orbit is dense. Since the transformation that sends x onto xx_0^{-1} is a homeomorphism, it sends the orbit of x_0, i.e., the sequence $\{c^n x_0\}$, onto a dense sequence; the image of that orbit, however, consists exactly of the powers of c.

Suppose, conversely, that $\{c^n\}$ is dense. If f is a character of X (i.e., a continuous homomorphism into the circle group), then $f(cx) = f(c)f(x)$, so that f is a proper vector of the unitary operator induced by T. Since the characters constitute a complete orthonormal set in L_2, every invariant function in L_2 can be expanded in terms of them. Since, for unitary operators, proper vectors with distinct proper values are orthogonal, every invariant function in L_2 (i.e., every proper vector with proper value 1) is a linear combination of the characters with proper value 1. The proof can be completed by showing that the only such character is the principal character. Indeed, if f is a character and $f(cx)=f(x)$ almost everywhere, then, by continuity, $f(cx)=f(x)$ everywhere, and therefore $f(c^n x)=f(x)$ everywhere. The result follows by setting x equal to 1.

Either the topological technique just described or the Fourier expansion method used for the circle group serves to show that a rotation on the torus $(T(x, y) = (bx, cy))$ is ergodic if and only if the coordinates of the multiplier, i.e., the numbers b and c, are integrally independent. This means that $b^n c^m = 1$ (for integers n and m) implies $n = m = 0$.

Can a linear transformation T of determinant 1 on a finite-dimensional real Euclidean space be ergodic? The answer turns out to be no. One way to see this is to apply the known proper-value theory of linear transformations on complex vector spaces; for this purpose the underlying space X must be complexified. Consider, in other words, the Cartesian product $X \times X$; define vector addition coordinatewise and define complex scalar multiplication by $(a+ib)(x, y) = (ax - by, bx + ay)$. The result is a complex vector space $\check{X}$ of complex dimension n (where n is the real dimension of X). The real dimension of $\check{X}$ is, of course, $2n$; the $2n$-dimensional real vector space $\check{X}$ includes X as an n-dimensional subspace. If $\check{T}$ is defined on $\check{X}$ by $\check{T}(x, y) = (Tx, Ty)$, then $\check{T}$ is a complex linear transformation with determinant 1. Let $c_1, \cdots, c_k$ be the distinct proper values of $\check{T}^*$ with multiplicities $n_1, \cdots, n_k$, and let $z_1, \cdots, z_k$ be corresponding non-zero proper vectors of $\check{T}^*$. Note that, by the definition of $\check{T}^*$, the vectors z_j are complex linear functionals on $\check{X}$. Since $X \subset \check{X}$, the functions z_j are defined on X; if $f(x) = \prod_{j=1}^{k} (z_j(x))^{n_j}$, then the function f is invariant under T. (Recall that the product of the proper values, each taken as often as its multiplicity, is the determinant.) The function f is not constant. Indeed, f vanishes on the union of the null-spaces of the z_j's, and nowhere else. Since the real and imaginary parts of each z_j are real linear functionals on X, the union of the null-spaces is a finite union of spaces of dimension less than n. Since f is continuous, these considerations imply that it is not an almost constant. (The complexification trick used here can be achieved with fewer concepts and many more indices by considering matrices instead of linear transformations.)

It would be of interest to know how far these facts can be extended.

Can an automorphism of a locally compact but non-compact group be an ergodic measure-preserving transformation? Nothing is known about this subject; only in the compact case has anything ever been done. I shall consider automorphisms of compact groups a little later.

For my next example I take T to be either the unilateral or the bilateral shift on the appropriate sequence space; I assert that in either case T is ergodic. Suppose indeed that E is a measurable invariant set. Since the measure is determined by its values on sets that depend on a finite number of coordinates only, there exists such a " finite-dimensional " set, say A, that is a close approximation to E. This means that the measure of the symmetric difference (or Boolean sum) $E + A$ is small, and it implies, in particular, that $m(E)$ is close to $m(A)$. Since A is determined by a finite set of coordinates, it follows that if n is sufficiently large, then the set $B = T^{-n}A$ is determined by a disjoint set of coordinates, and therefore that $m(A \cap B) = m(A)m(B)$. Since T and all its powers are measure-preserving and since E is invariant, the smallness of $m(E + A)$ implies that of $m(E + B)$. It follows that $m(E + (A \cap B))$ is small also and hence that $m(A), m(B)$, and $m(A)m(B)$ are all close to $m(E)$; in other words $m(E)$ is nearly equal to its own square. Since the degree of approximation is arbitrary, it follows that $m(E) = (m(E))^2$, as desired. From this it follows that both the doubling transformation ($Tx = 2x \pmod 1$) and the baker's transformation are ergodic.

With the exception of the unit translation on the space of integers, all our examples of ergodic transformations so far have acted on spaces of finite measure. It is a non-trivial task to construct an example of an ergodic transformation on a non-atomic space of infinite measure. (Non-atomic means that every measurable set of positive measure includes a measurable subset of smaller positive measure.) I shall show how to construct such a transformation on the real line. Actually the most convenient representation of the space on which the transformation acts is not the real line but a certain collection of segments in the plane; it will be clear, however, that if those segments were strung together, they would add up to the real line. The method has wide applications in the construction of examples.

Let $\{a_n\}$ be a decreasing sequence of positive numbers with $a_0 = 1$, and let T_0 be an invertible ergodic measure-preserving transformation on the half-open unit interval. Let X_n be a half-open interval of length a_n, located in the plane parallel to the horizontal axis; let X be the union of all the X_n. (If $\sum_n a_n$ diverges, then X is in an obvious way measure-theoretically isomorphic to a half-line, and hence to a line.) The transformation T on X moves every point up by one unit whenever possible (i.e., $T(x, y) = (x, y+1)$); when that is no longer possible (i.e., when $y = n$ and $x \geqq a_{n+1}$), then $T(x, y) = (T_0 x, 0)$. It is clear that T is an invertible measure-preserving transformation on X. Suppose now that a measurable set E is invariant under T. If E_0 is the intersection of E with the base interval (i.e., with the unit interval on the horizontal axis), then E consists exactly of the points whose first coordinate is in E_0. Since E_0 is invariant under T_0, either E_0 or its relative complement in the base interval has measure zero; it follows, as asserted, that T is ergodic.

CONSEQUENCES OF ERGODICITY

For ergodic transformations the statement of the ergodic theorem can be strengthened by adjoining to it a description of the limit function. Precisely speaking, if T is an ergodic measure-preserving transformation, if f is an integrable function, and if $f^*(x) = \lim \frac{1}{n} \sum_{j=0}^{n-1} f(T^j x)$, then $f^*(x)$ is equal almost everywhere to a constant. If the measure of the underlying space is infinite, the constant is 0. The reason is that f^* is integrable and 0 is the only integrable constant. (This proves again that for spaces of infinite measure we have no right to expect that $\int f(x)dx = \int f^*(x)dx$.) If the measure of the space is finite, then the equality of the integrals of f and f^* shows that $f^* = \frac{1}{m(X)} \int f(x)dx$. In other words, for ergodic measure-preserving transformations on spaces of finite measure the phase (or space) mean $\frac{1}{m(X)} \int f(x)dx$ is equal almost everywhere to the time mean $\lim \frac{1}{n} \sum_{j=0}^{n-1} f(T^j x)$. This assertion, of great significance in the physical aspects of the theory, is sometimes (incorrectly) identified with the ergodic theorem. An open question of both mathematical and physical interest is to find usable (i.e., easily verifiable) conditions on a transformation that imply ergodicity.

For spaces of finite measure the condition that the time mean of every integrable function be a constant is sufficient for ergodicity as well as necessary. To prove this, it is sufficient to prove that every invariant function in L_1 is a constant, and that follows from the fact that an invariant function is its own time mean. On a space of infinite measure it can happen that all time means are constants even for transformations that are not ergodic; for an example let X be the set of all lattice points in the plane and define T by $T(x, y) = (x+1, y)$. In this situation, the constant 0 is the only invariant function in L_1.

31

For ergodic transformations, spaces of finite measure, and non-negative measurable functions, the converse of the ergodic theorem holds, in the following sense: if $\dfrac{1}{n}\sum\limits_{j=0}^{n-1} f(T^jx)$ tends to a finite limit almost everywhere, then f is integrable. The proof begins with the observation that the limit is equal almost everywhere to a constant c. If f_k is the function obtained from f by truncating it at k (i.e., $f_k(x)=f(x)$ when $f(x)\leq k$ and $f_k(x)=k$ when $f(x)>k$), then f_k is bounded and therefore integrable. The theorem on the integration of monotone sequences implies that $\int f_k(x)dx$ tends to $\int f(x)dx$; the integrability of f can be proved by showing the boundedness of the sequence $\{\int f_k(x)\,dx\}$. Since $f_k\leq f$, it follows that $\dfrac{1}{n}\sum\limits_{j=0}^{n-1} f_k(T^jx)\leq\dfrac{1}{n}\sum\limits_{j=0}^{n-1} f(T^jx)$ and therefore that $f_k{}^*\leq c$ almost everywhere. Consequently $\int f_k{}^*(x)dx\leq cm(X)$ for all k, and, therefore, the integrals $\int f_k(x)dx$ have the same bound.

The preceding result extends immediately to semi-integrable real-valued measurable functions, i.e., to measurable functions f such that either the positive part or the negative part of f is integrable. The fact that the result does not extend to all real-valued measurable functions is established by the following example constructed by M. Gerstenhaber. The space is of the step-ladder type used before in the construction of an ergodic transformation on the line; the transformation T is the dilation of an ergodic transformation T_0 on the base interval, just as in that example. It follows that T itself is ergodic. The auxiliary numbers a_n (i.e., the lengths of the intervals X_n that constitute the space X) are selected so that (i) $a_{2n-1}=a_{2n}$, $n=1, 2, 3,\cdots$, (ii) $\sum_n a_{2n}$ converges, and (iii) $\sum_n a_{2n}\sqrt{n}$ diverges. This can all be done by, for instance, setting a_{2n} equal to $n^{-3/2}$. It follows from (ii) that $m(X)$ is finite. If $f(x)=\sqrt{n}$ when $x\in X_{2n}$ and $f(x)=-\sqrt{n}$ when $x\in X_{2n-1}$, then it follows from (iii) that f is not integrable; in fact neither the positive nor the negative part of f is integrable. Since, by (i), in a sum of the form $\sum\limits_{j=0}^{n-1} f(T^jx)$ all but, at worst, the two extreme terms cancel out, and the absolute values of those terms are dominated by $\sqrt{n}$, it follows that $\lim \dfrac{1}{n}\sum\limits_{j=0}^{n-1} f(T^j x)$ exists and is, in fact, equal to 0 for all x.

For an ergodic transformation T on a space of measure 1 Kac made an interesting addition to the recurrence theorem (Bull. A.M.S., 1947, p. 1006). If E is a measurable set of positive measure, and if $n(x)$ denotes, for each x in E, the least positive integer such that $T^{n(x)} \in E$, then $n(x)$ is defined almost everywhere in E. It is easy to verify that $n(x)$ depends measurably on x; Kac's theorem is that $\int_E n(x)dx = 1$. (The proof is not difficult, but it involves some combinatoric trickery; I omit it.) If the conclusion is expressed in the form $\dfrac{1}{m(E)}\int_E n(x)dx = \dfrac{1}{m(E)}$, then it can be given the following verbal statement: the average length of time that it takes a point of E to return to E is the reciprocal of the measure of E. For non-ergodic transformations this is false.

Suppose again that T is an ergodic measure-preserving transformation on a space X of finite measure, and let F and G be any two measurable subsets of X, with respective characteristic functions f and g. The fact that $f^*(x) = \dfrac{1}{m(X)}\int f(x)dx$, i.e., $f^*(x) = m(F)/m(X)$ almost everywhere, can be expressed as follows: the mean time of sojourn in F of almost every orbit is proportional to $m(F)$. Since $\lim \dfrac{1}{n}\sum\limits_{j=0}^{n-1} f(T^j x)g(x) = f^*(x)g(x)$, it follows from the bounded convergence theorem that $\lim \dfrac{1}{n}\sum\limits_{j=0}^{n-1} m(T^{-j}F \cap G)$ $= m(F)m(G)/m(X)$. If, for every measurable set E, we interpret $m(E)/m(X)$ to be the probability that a point belong to E, the last result can be expressed as follows: the probability that an iterated transform of a point of G belong to F converges in the sense of Cesàro to the product of the probabilities of F and of G. In other words: a moving set G tends on the average to become stochastically independent of each fixed set F.

Ergodicity has a strong and surprising influence on the spectral structure of the induced unitary operator; the basic known facts can be summarized as follows.

PROPER VALUE THEOREM. *An invertible measure-preserving transformation T on a space of finite measure is ergodic if and only if the number* 1 *is a simple proper value of the induced unitary operator U. If T is ergodic, then the absolute value of every proper function of U is constant, every proper value is simple, and the set of all proper values of U is a subgroup of the circle group.*

PROOF. Since the space has finite measure, every constant function f is in L_2; since $Uf=f$, the number 1 is always a proper value of U. Since the set of all constant functions is a one-dimensional subspace of L_2, and since T is ergodic if and only if the only invariant functions in L_2 are the constants, the first assertion of the theorem is proved. (Recall that a function in L_2 is invariant if and only if it is a proper function of U corresponding to the proper value 1).

Since U is unitary, every proper value of U has absolute value 1. It follows that if f is a proper function with proper value c, i.e., $f(Tx)=cf(x)$ almost everywhere, then $|f|$ is invariant; the ergodicity of T implies then that $|f|$ is a constant. If both f and g are proper functions with proper value c, then f/g is an invariant function, so that g is a constant multiple of f. (Note that since $|g|$ is a non-zero constant, f/g makes sense.) This proves the simplicity of each proper value. If, finally, b and c are proper values of U, with corresponding proper functions f and g, then f/g is a proper function of U with proper value b/c; this proves that the proper values of U form a group.

I conclude this preliminary study of ergodicity by describing a particular example of a non-ergodic transformation and commenting on the conjecture that it suggests. The space X on which the transformation acts is the unit square, or rather, since operations modulo 1 will be relevant, the torus; the transformation is defined by $T(x, y)=(x, y+x)$ (mod 1). If $f(x, y)=g(x)$, where g is a measurable function on the unit interval, then f is invariant under T; the abundance of such invariant functions shows that T is not ergodic. For each fixed x_0, the vertical segment over x_0 (i.e., the set of all (x_0, y)) is invariant under T. The restriction of T to such a segment is a measure-preserving transformation; for almost all such segments that restriction is ergodic. (In fact the restriction is ergodic for all but the countably many segments for which x_0 is rational.) Thus, in an intuitively obvious sense, the original transformation T is a direct sum (direct integral) of ergodic transformations; the given decomposable transformation has been decomposed into indecomposable pieces. It is natural to conjecture that this situation is typical, and, in a certain sense, the conjecture is true. Since, however, the proof is quite delicate, and since, worse yet, the result does not turn out to be useful, I shall omit the detailed discussion of the decomposition theorem here. I mention the theorem for its heuristic value. The right theorems in this subject can usually be guessed by pretending that the decomposition theorem is available; their proofs can then be obtained directly, i.e., without the delicate apparatus of the decomposition theorem.

As an example of the sort of result the decomposition theorem suggests, I consider the problem of uniqueness of an invariant measure. Suppose first that T is invertible and ergodic; what can be said about a measure p with the same domain as m, equivalent to m (i.e., $p(E) = 0$ if and only if $m(E) = 0$), and invariant under T? The answer is that p is a constant multiple of m. For the proof, apply the Radon-Nikodym theorem to write $p(E) = \int_E f(x) dm(x)$. Since $p(TE) = \int_{TE} f(x) dm(x)$, the invariance of p and an application of the change of variables that replaces x by Tx yield $\int_E f(x) dm(x) = \int_E f(Tx) dm(x)$ for every measurable set E. This implies that f is a constant almost everywhere; since m is absolutely continuous with respect to p, the constant is not zero. In case $m(X)$ is finite, the result may be expressed as follows: if T is ergodic, then there is a unique measure p that is equivalent to m and invariant under T, and that assumes a prescribed value on X. A glance at the ergodic pieces of a non-ergodic transformation suggests the generalization: if T is a measure-preserving transformation on a space of finite measure, and if m_0 is a prescribed measure on the algebra of all invariant sets that is equivalent to m on that algebra, then there exists a unique finite invariant measure p that is defined for all measurable sets, equivalent to m, and equal to m_0 on the invariant sets. This generalization is correct. The proof is similar to the Radon-Nikodym technique used before; the details are omitted.

MIXING

Ergodic theory has been developed most extensively for invertible measure-preserving transformations on a space of finite measure. In order to discuss that development, I shall in the sequel restrict my attention to that case. In the absence of any statement to the contrary, from now on the word transformation will mean invertible measure-preserving transformation, and the measure m of the underlying space X will be assumed to be normalized so that $m(X) = 1$.

We have seen that if a transformation T is ergodic, then $m(T^{-n}F \cap G)$ converges in the sense of Cesàro to $m(F)m(G)$. The validity of this condition for all F and G is, in fact, equivalent to ergodicity. To prove this, suppose that E is a measurable invariant set, and take both F and G equal to E. It follows that $m(E) = (m(E))^2$, and hence that $m(E)$ is either 0 or 1. This convergence condition for ergodicity has a functional form also: T is ergodic if and only if $\int f(T^n x)g(x)dx$ converges Cesàro to $\int f(x)dx \cdot \int g(x)dx$ whenever f and g are in L_2. It is sufficient to prove the only if. For this purpose, observe that if $f \in L_2$, then, by the mean ergodic theorem, $\frac{1}{n}\sum_{j=0}^{n-1} f(T^j x)$ converges in the mean of order two to $f^*(x)$; it follows from the Schwarz inequality that $\frac{1}{n}\sum_{j=0}^{n-1} f(T^j x)g(x)$ converges in the mean of order one to $f^*(x)g(x)$. Since L_1 convergence implies term-by-term integrability, and since, by ergodicity, $f^*(x) = \int f(x)dx$ almost everywhere, the proof is complete.

The Cesàro convergence condition has a natural intuitive interpretation. We may visualize the transformation T as a particular way of stirring the contents of a vessel (of total volume 1) full of an incompressible fluid, which may be thought of as 90 per cent gin and 10 per cent vermouth. If G is the region originally occupied by the vermouth, then, for any part F of the vessel, the relative amount of vermouth in F, after n repetitions of the act of stirring, is given by $m(T^{-n}F \cap G)/m(G)$. The ergodicity of T implies therefore that on the average this relative amount is exactly equal to 10 per cent. In general, in physical situations like this one, one expects to be justified in making a much stronger statement, namely that, after the liquid has been stirred sufficiently often, every part F of the container will contain approximately 10 per cent vermouth. In mathematical language this expectation amounts to replacing Cesàro convergence by ordinary convergence, i.e., to the condition $\lim m(T^{-n}F \cap G)$ $= m(F)m(G)$. If a transformation T satisfies this condition for every pair F and G of measurable sets, it is called mixing, or, in distinction from a related but slightly weaker concept, strongly mixing.

Mixing also has a functional form: T is mixing if and only if $(U^n f, g)$ converges to $(f, 1)(1, g)$ whenever f and g are in L_2. (The operator U is, of course, the unitary operator induced by T.) If f and g are the characteristic functions of F and G respectively, then the functional form just stated reduces to the set-theoretic form that appears in the definition. The general functional form is derived from the set-theoretic form by a double approximation process. I argue first that, for each fixed characteristic function g, the result is valid for all simple functions f, and therefore, by L_2 approximation, for all functions f in L_2; second, I hold f fixed in L_2 and argue similarly about g. Since $(f, 1)(1, g) = ((f, 1)1, g)$, the result can be expressed as follows: T is mixing if and only if the powers of U converge in the weak operator topology to the operator P defined by $Pf = (f, 1)1$. The operator P is the projection on the space of constant functions.

A group rotation (e.g., the transformation T defined on the circle group by $Tx = cx$) is not mixing. Indeed, if $f(x) = x$, then $Uf = cf$ and therefore $U^n f = c^n f$. It follows that $(U^n f, f) = c^n$, whereas $(f, 1)(1, f) = 0$. The easiest example of a mixing transformation is the baker's transformation, i.e., the bilateral shift. To prove this, given F and G, find finite-dimensional sets A and B that approximate F and G. Since $m(T^{-n}A \cap B)$ $= m(A)m(B)$ for large n, it follows that $m(T^{-n}F \cap G)$ is close to $m(F)m(G)$ for large n.

Between ergodicity and mixing there is room for another concept—the concept of weak mixing. This apparently artificial concept is of great technical significance. A transformation T is, by definition, weakly mixing if

$$\lim \frac{1}{n}\sum_{j=0}^{n-1} |m(T^{-j}F\cap G)-m(F)m(G)| = 0$$

for every pair F and G of measurable sets. The functional form of this condition, easily proved to be equivalent to the set-theoretic form, is that

$$\lim \frac{1}{n}\sum_{j=0}^{n-1} |(U^{j}f,g)-(f,1)(1,g)| = 0$$

whenever f and g are in L_2. In technical language, the definition of weak mixing substitutes strong Cesàro convergence for the Cesàro convergence that takes place when T is ergodic and for the ordinary convergence that occurs in the definition of strong mixing.

There are some amusing analytic exercises connected with the type of convergence we are now considering. In order to justify the statement I made about the place of weak mixing (between ergodicity and strong mixing), it has to be proved that if $\{a_n\}$ is a sequence of complex numbers such that $\lim a_n=a$, then $\lim \frac{1}{n}\sum_{j=0}^{n-1} |a_j-a| =0$, and also that if $\lim \frac{1}{n}\sum_{j=0}^{n-1} |a_j-a| =0$, then $\lim \frac{1}{n}\sum_{j=0}^{n-1} a_j=a$. These things are easy. An only slightly harder but considerably more interesting fact is this: for bounded sequences $\{a_n\}$, a necessary and sufficient condition that $\lim \frac{1}{n}\sum_{j=0}^{n-1} |a_j-a|$ $=0$ is that there exist a set J of positive integers such that J has density zero and such that if n is restricted to be outside J, then $\lim a_n=a$. (To say that J has density zero means that the ratio of the number of those integers between 0 and $n-1$ that belong to J to the total number, i.e., to n, tends to 0 as n tends to infinity.) If, in the gin and vermouth example, ergodicity is expressed by saying that on the average F has 10 per cent vermouth, and if strong mixing is expressed by saying that after a while F will have 10 per cent vermouth in it, then weak mixing can be expressed by saying that after a while F will have 10 per cent vermouth in it, with the exception of a few rare instants during which it may be either too strong or too sweet.

Weak mixing is not just an analytic artificiality; its importance comes from the fact that it is equivalent to some rather natural geometric and functional conditions. In order to state the result, I define two new concepts. I shall say that a measure-preserving transformation T has continuous spectrum if the only proper value of the induced unitary operator U is the number 1 and if that proper value is simple. The expression (T has continuous spectrum) is a little too elliptical. Since the constant functions are always invariant under T, the number 1 is always a proper value of U, so that U always has some point spectrum. According to the usage I just established, T has continuous spectrum if this minimal point spectrum is all that appears in U. The second concept is the Cartesian square $\tilde{T}$ of a measure-preserving transformation T on a space X; the transformation $\tilde{T}$ is defined on the space $\tilde{X}$, the Cartesian product of X with itself, by $\tilde{T}(x, y) = (Tx, Ty)$. The unitary operator induced by $\tilde{T}$ will be denoted by $\tilde{U}$.

MIXING THEOREM.   *A transformation T is weakly mixing if and only if it has continuous spectrum, or, alternatively, if and only if its Cartesian square is ergodic.*

PROOF.   Suppose first that T is weakly mixing. To prove that $\tilde{T}$ is ergodic, it is sufficient to prove that $\tilde{m}(\tilde{T}^{-n}A \cap B)$ converges (Cesàro) to $\tilde{m}(A)\tilde{m}(B)$ whenever A and B are measurable rectangles in X, where $\tilde{m}$, of course, is the product measure in $\tilde{X}$. If $A = C \times D$ and $B = F \times G$, where C, D, F, and G are measurable subsets of X, then $\tilde{m}(\tilde{T}^{-n}A \cap B) = m(T^{-n}C \cap F)m(T^{-n}D \cap G)$. Since, by assumption, $m(T^{-n}C \cap F)$ converges (strong Cesàro) to $m(C)m(F)$, and, similarly, $m(T^{-n}D \cap G)$ converges (strong Cesàro) to $m(D)m(G)$, and since $\tilde{m}(A) = m(C)m(D)$ and $\tilde{m}(B) = m(F)m(G)$, it is sufficient to prove the following analytic lemma: if $\{a_n\}$ and $\{b_n\}$ are bounded sequences such that a_n and b_n converge (strong Cesàro) to a and b, respectively, then $a_n b_n$ converges (Cesàro) to ab. This is true with room to spare; in fact $a_n b_n$ converges (strong Cesàro) to ab. The result is an immediate consequence of the characterization of strong Cesàro convergence in terms of convergence outside a set of density zero, together with the fact that the union of two sets of density zero is another such set.

Suppose next that $\tilde{T}$ is ergodic. If f is a proper function of U, say $Uf = cf$, write $\tilde{f}(x, y) = f(x)\overline{f(y)}$. It follows that $\tilde{U}\tilde{f} = \tilde{f}$, and hence, since $\tilde{T}$ is ergodic, that $\tilde{f}$ is a constant. It is now clear that f must be a constant, and that c must be equal to 1.

Suppose finally that T has continuous spectrum; it is to be proved that T is weakly mixing. This is the deepest part of the proof; I need to make use of the spectral theorem for unitary operators and of some auxiliary analytic technique. I remark first of all that it is sufficient to prove that $\frac{1}{n}\sum_{j=0}^{n-1}|(U^jf,f)-(f,1)(1,f)|$ tends to 0; the standard polarization trick then implies the general result (i.e., the result with g in place of the second f). If f is a constant c, then $(U^jf,f)=|c|^2$ and $(f,1)(1,f)=|c|^2$; it is, therefore, sufficient to prove the result under the added assumption that $(f, 1)=0$. The next remark is that it is sufficient to prove that $\frac{1}{n}\sum_{j=0}^{n-1}|(U^jf,f)|^2$ tends to 0. (This is elementary analysis: for bounded sequences strong Cesàro convergence is equivalent to quadratic strong Cesàro convergence. One way to see this is by reference to the characterization in terms of sets of density zero; direct proofs are also easy to construct.) If E is the spectral measure of U, then $(U^jf,f)=\int x^j d(E(x)f,f)$. Since, by assumption, f is orthogonal to all proper functions of U, it follows that the integrating measure, i.e., the measure p defined for all Borel subsets M of the circle by $p(M)=(E(M)f,f)$, gives measure zero to each one-point set. In other words, p is non-atomic. The desideratum has now been reduced to the following assertion: if p is a non-atomic measure in the circle, then $\frac{1}{n}\sum_{j=0}^{n-1}\left|\int x^j dp(x)\right|^2$ tends to 0. The next step is to replace $\left|\int x^j dp(x)\right|^2$ by $\int x^j dp(x)\cdot\int \bar{y}^j dp(y)$, combine the integrals to a double integral, and bring the summation inside. The desired result becomes then

$$\lim \int\int \frac{1}{n}\sum_{j=0}^{n-1}(x\bar{y})^j dp(x)dp(y)=0.$$

The non-atomicity of p implies that the diagonal of the torus (the Cartesian product of the circle with itself) has measure zero with respect to the product measure. It follows that the integrand is equal almost everywhere to

$$(1 - (x\bar{y})^n)/n(1 - x\bar{y}).$$

Since, therefore, the integrand tends to 0 almost everywhere, and since the integrand is bounded (in fact, it is bounded by 1—cf. the sum representation), the proof can be completed by reference to the Lebesgue bounded convergence theorem.

MEASURE ALGEBRAS

Ergodic theory can be studied on three different levels; they are adequately described by the words algebraic, geometric, and analytic. The geometric level is the one that has been receiving the most attention so far; it is the one that is concerned with transformations on a measure space. The analytic level has also been mentioned; it is the one that studies the linear operators induced by a transformation on the various L_p spaces. The algebraic aspect of the theory is to my mind the cleanest and the most natural; it studies the group of automorphisms of certain Boolean algebras.

Many of the difficulties of measure theory and all the pathology of the subject arise from the existence of sets of measure zero. The algebraic treatment gets rid of this source of unpleasantness by refusing to consider sets at all; it considers sets modulo sets of measure zero instead. Suppose, to be specific, that X is a measure space with a normalized measure m, and let B be the set of all equivalence classes of measurable sets, where two measurable sets E and F are called equivalent if and only if their symmetric difference $E+F$ has measure zero. The set B is a Boolean algebra under the natural Boolean operations. Indeed if (E_1, E_2) and (F_1, F_2) are pairs of equivalent sets, then $E_1 \cup F_1$ is equivalent to $E_2 \cup F_2$; it follows that the " union " of two equivalence classes can be unambiguously defined by selecting representatives from each and forming the equivalence class of their union. The same is true for intersections and complements, and, since a measure is countably additive, it is true for countable unions and countable intersections also. The zero element of the Boolean algebra B is the class of all sets of measure zero; the unit element is the class of all sets of measure one. Since $m(E+F)=0$ implies that $m(E)=m(F)$, the function m may be considered as defined on B; it is in an obvious sense a measure on B. The only element of B that has measure zero is the zero element; similarly, the only element of B that has measure one is the unit element. A structure such as (B, m), i.e., a Boolean sigma-algebra equipped with a strictly positive normalized measure, is called a measure algebra. The concept of a measure algebra is the algebraic substitute for the geometric concept of a measure space.

A measure-preserving transformation T on X induces in a natural way a mapping of B into itself. The image of an equivalence class under this mapping is defined by selecting a representative E and forming the equivalence class of $T^{-1}E$; the measure-preserving character of T implies that the image class is unambiguously determined by this process and that the measure of the image class is the same as the measure of the original one. The measure-preserving mapping of B into itself thus determined will be denoted by T^{-1}. The mapping T^{-1} on B preserves all the Boolean operations (including the countably infinite ones); it is an isomorphism of B into (but not necessarily onto) itself. A necessary and sufficient condition that T^{-1} be an automorphism of B is that the transformation T be invertible.

Several of the concepts and results of ergodic theory extend easily to monomorphisms of a measure algebra, and, in fact, the entire subject could be treated within that framework. Instead of doing that, I shall try to have the best of both worlds; in the sequel both algebraic and geometric concepts will be used, the choice in each case being a matter of convenience.

If B is the measure algebra associated with a measure space X, then every invertible measure-preserving transformation on X is (induces) an automorphism of B. Is it true, conversely, that every automorphism of B is so induced? The answer is no, in general. There exist some highly pathological measure spaces that are in a certain vague sense absolutely non-measurable; one way that the pathology shows up is that they do not possess sufficiently many measure-preserving transformations to induce all the desired Boolean automorphisms. Since it can be argued that sets of measure zero are worthless, not only from the algebraic but also from the physical point of view, and since every measure algebra can be represented as the algebra associated with a non-pathological measure space, the poverty of some measure spaces may be safely ignored. The fact that the associated measure algebra might have more automorphisms than the original measure space has transformations is an advantage of the algebra, not a hindrance.

The moralizing tone of the preceding discussion can be made somewhat more specific by discussing the question: when are two invertible measure-preserving transformations S and T essentially the same? There are three possible answers. If S and T are regarded as transformations on a measure space X, then the proper answer is that there exists an invertible measure-preserving transformation Q on X such that $S = Q^{-1}TQ$; in this case S and T will be called (geometrically) similar. If S and T are regarded as automorphisms of a measure algebra B, then the proper answer is that there exists an automorphism Q of that algebra such that $S = Q^{-1}TQ$; in this case S and T will be called (algebraically) conjugate. If, finally, S and T are regarded as unitary operators on a Hilbert space H, then the proper answer is that there exists a unitary operator Q on H such that $S = Q^{-1}TQ$; in this case S and T will be called (spectrally) equivalent. For later purposes it is important to observe that similarity, conjugacy, and equivalence can also be defined for pairs of transformations that do not act on the same domain; in that case the implementing transformation Q will map one of the two domains onto the other. In an obvious sense similarity implies conjugacy, and conjugacy implies equivalence. The converse is false for both these implications. Equivalence does not imply conjugacy for a sound and interesting algebraic reason; we shall presently look into the situation. Conjugacy does imply similarity in all decent measure spaces but not in the indecent ones; we hereby abandon all interest in similarity.

I conclude this section by discussing the relation between conjugacy and equivalence. Let B be the measure algebra of a measure space with measure m. If T is an automorphism of B, then the induced unitary operator U on L_2 preserves more than the norm and the linear structure of L_2. The more comes from the fact that the elements of L_2 are not merely abstract vectors; being functions (or, rather, equivalence classes of functions), they have multiplicative properties also. If, for safety, we stick to bounded functions, then multiplication does not lead out of L_2, and the unitary operator U preserves products. This condition on U turns out to be sufficient as well as necessary in order that U be induced by an automorphism.

MULTIPLICATION THEOREM. *A unitary operator U on L_2 is induced by an automorphism T of B if and only if both U and U^{-1} send every bounded function onto a bounded function and $U(fg) = (Uf)(Ug)$ whenever f and g are bounded functions.*

PROOF. It is sufficient to prove that if U is multiplicative and boundedness-preserving, then it is induced. If f is a characteristic function, corresponding to an element F of B, then $f^2 = f$, and therefore $(Uf)^2 = Uf$. This proves that Uf is a characteristic function also, corresponding to G, say. If we denote G by TF, then T is a mapping of B into itself. The fact that U is onto implies that T is onto, and the fact that U has no null space implies that if $TF = 0$, then $F = 0$. The proof can be completed by showing that T is a measure-preserving sigma-homomorphism. The measure-preserving character of T follows from the norm-preserving character of U; recall that $m(F) = \|f\|^2$. The fact that T preserves intersections is immediate from the multiplicativity of U. The preservation of unions follows from the fact that if f and g are characteristic functions, corresponding to F and G, say, then the characteristic function of $F \cup G$ is $f + g - fg$. The preservation of countable unions follows (after an obvious induction argument) from the continuity of U.

The relation between conjugacy and equivalence is now clear. If S and T are automorphisms, then a necessary and sufficient conditiont that they be conjugate is that they be equivalent and that the unitary operator that implements the equivalence can be chosen to be multiplicative.

DISCRETE SPECTRUM

A transformation T is said to have discrete spectrum (or pure point spectrum) if there is a basis $\{f_j\}$ of L_2 (i.e., a complete orthonormal set) each term of which is a proper vector of the induced unitary operator U.

DISCRETE SPECTRUM THEOREM. *Two ergodic transformations with discrete spectrum are conjugate if and only if their induced unitary operators are equivalent.*

PROOF. It is sufficient to prove that equivalence implies conjugacy. Let the given transformations be S and T, and let the induced unitary operators be U and V. Let C be the set of all proper values of U; since U and V are equivalent, C is also the set of all proper values of V. To each c in C there corresponds a proper vector f_c of U. The proper value theorem implies that $|f_c|$ is a constant; there is no loss of generality in assuming that $|f_c| = 1$. The proper value theorem implies also (since S is ergodic) that f_c is now uniquely determined to within a constant factor of absolute value 1. The fact that U has discrete spectrum implies that the family $\{f_c\}$ is a basis for L_2.

If a and b are in C, then $f_a f_b$ is a proper vector of U with proper value ab; it follows that there exists a constant $r(a, b)$ of absolute value 1 such that $f_a f_b = r(a, b) f_{ab}$. I assert that there exists a homorphism p from the group of all functions of constant absolute value 1 onto the circle such that each constant function is carried onto itself. If the functions in question were really functions, instead of equivalence classes of functions modulo sets of measure zero, this would be trivial: evaluation at an arbitrary point would be such a homomorphism. Since this technique is not available, a slightly more sophisticated group-theoretic technique is needed. Let us take the result granted for a moment. If $s_a = p(f_a)$, then an application of p to the equation that determines the multiplier constant r yields that $s_a s_b = r(a, b) s_{ab}$. It follows that if $\bar{f}_c = \bar{s}_c f_c$, then the mapping that assigns to each c in C the function $\bar{f}_c$ is a homomorphism. In other words, there is no loss of generality in assuming that $f_a f_b = f_{ab}$ whenever a and b are in C. Similarly, we can find for each c in C a function g_c of constant absolute value 1 so that g_c is a proper vector of V with proper value c, so that the family $\{g_c\}$ is a basis for L_2, and so that $g_a g_b = g_{ab}$ whenever a and b are in C.

It is now time to settle the group-theoretic lemma needed in this proof. Since the elements of the circle group may be identified with the corresponding constant functions, and since the circle group is divisible (i.e., every element has roots of all orders), the lemma can be formulated as follows : if H is an abelian group and K is a divisible subgroup, then K is a retract of H, i.e., there exists a homomorphism from H onto K that is the identity on K. (For appearances of this lemma in other contexts, see Weil, L'intégration dans les groupes topologiques, 1940, p. 94, or Kaplansky, Infinite abelian groups, 1954, p. 8.) For the proof, order the retractions onto K from supergroups of K in H by extension, and Zornify. If p is a maximal such retraction, then its domain L must be H. Suppose, indeed, that g is in $H - L$, and let M be the subgroup generated by L and g. Every element of M has the form hg^j with h in L and j an integer. If no positive power of g is in L, then this representation is unique ; if n is the least positive integer such that g^n is in L, then there is a unique such representation with $0 \leqq j < n$. In the first case write $q(hg^j) = p(h)$; in the second case let g_0 be an n-th root of $p(g^n)$ in K and write $q(hg^j) = p(h)g_0^j$ ($0 \leqq j < n$). In either case q is a retraction of M onto K, and q is an extension of p ; this contradicts the maximality of p.

To complete the proof of the discrete spectrum theorem, let W be the unitary operator such that $Wg_c = f_c$ for all c in C. I assert that W is multiplicative, i.e., if g and h are bounded functions, then $W(gh) = (Wg)(Wh)$. If $g = g_a$ and $h = g_b$, with a and b in C, this is obvious from the definition of W and the multiplicative properties of the families $\{f_c\}$ and $\{g_c\}$. By linearity the result extends to finite linear combinations of g_c's. The passage to the limit (first with the bounded function h being held fixed, and then ditto with g) causes no difficulty. The multiplication theorem implies that W is induced by an automorphism of the measure algebra. The proof is completed by the following elementary computation : $W^{-1}UWg_c = W^{-1}Uf_c = cW^{-1}f_c = cg_c$; this shows that $W^{-1}UW = V$.

The discrete spectrum theorem can be used to answer virtually all questions about ergodic transformations with discrete spectrum; the following results indicate the general method.

REPRESENTATION THEOREM. *An ergodic measure-preserving transformation with discrete spectrum is conjugate to a rotation on a compact abelian group.*

PROOF. Let C be the spectrum (i.e., the set of all proper values) of the given transformation; let X be the character group of C. If $z(c) = c$ for every c in C, then z is an element of X. The rotation T on X, defined by $Tx = zx$, is a measure-preserving transformation with discrete spectrum, and, moreover, its spectrum is exactly C. The discreteness of the spectrum follows from the properties of the characters of X. They form a complete orthonormal set in the L_2 space over X, and, if f_0 is one of them, then $f_0(zx) = f_0(z)f_0(x)$, so that f_0 is a proper function with proper value $f_0(z)$, This argument shows also that the spectrum of T is the set of all f_0 (z)'s, each occurring with multiplicity equal to the number of characters f of X for which $f(z) = f_0(z)$. If to each c in C we make correspond the function f_c on X defined by $f_c(x) = x(c)$, then this correspondence is an isomorphism from C onto the character group of X. Since $f_c(z) = c$ for each c, it follows, as asserted, that the spectrum of T is C. Since the same relation shows that each element of C has multiplicity one in the spectrum of T, the rotation T is ergodic. The representation theorem is now an immediate consequence of the discrete spectrum theorem.

The representation theorem and its proof have several interesting corollaries.

COROLLARY 1. *Every subgroup of the circle is the spectrum of an ergodic measure-preserving transformation with discrete spectrum.*

PROOF. The construction in the proof of the representation theorem used the spectrum of C only.

COROLLARY 2. *If T is an ergodic measure-preserving transformation with discrete spectrum, then T is conjugate to the product of two involutions (where an involution is a transformation S such that S^2 is the identity).*

PROOF. By the representation theorem, I may assume that T is a rotation, say $Tx = cx$, on a compact abelian group X. If b is in X and $Rx = bx^{-1}$, then R is an involution; if $Sx = RTx$, then $Sx = bc^{-1}x^{-1}$, so that

S is an involution also, and, clearly, $RS = T$. This concludes the proof.

COROLLARY 3. *An ergodic measure-preserving transformation with discrete spectrum is conjugate to its own inverse.*

PROOF. Immediate from Corollary 2; the proof even shows that the conjugacy can be implemented by an involution.

The simplest examples of measure-preserving transformations are the ones with a purely atomic domain. Suppose, for instance, that $Tn = n+1$ on the discrete space of integers. If $Pn = 1-n$ and $Qn = -n$ then $T = PQ$. Since these equations continue to make sense even when they are reduced by an arbitrary integral modulus, it follows that every cyclic (i.e., ergodic!) permutation of a finite set is a product of two involutions. Since every finite permutation is a product of disjoint cycles, the result remains true for all (i.e., not necessarily ergodic) finite permutations. It follows that every finite permutation is similar to its own inverse. This latter fact is obvious for another reason : since the similarity class of a permutation is determined by the cardinal numbers associated with its cycle decomposition (i.e., by the numbers of cycles of each length), the fact that a permutation is similar to its inverse is a consequence of the fact that a permutation and its inverse have the same kind of cycle decomposition. I mention these elementary facts because they constitute the discrete skeleton that supports the continuous generalizations obtained above as corollaries of the representation theorem.

When does a measure-preserving transformation T have a square root, i.e., when does there exist a measure-preserving transformation S such that $S^2 = T$? Since the only solved fragment of this problem concerns transformations with discrete spectrum, this is the proper place to report the solution.

SQUARE ROOT THEOREM. *If T is an ergodic measure-preserving transformation with discrete spectrum on a space of finite measure, then a necessary and sufficient condition that T be conjugate to a square is that -1 be absent from its proper values.*

PROOF OF NECESSITY. (This part of the proof does not need discrete spectrum.) Observe first that every set (or function) invariant under a transformation S is automatically invariant under S^2. Hence it is sufficient to prove that if S is an ergodic transformation with an ergodic square, and if f is a function in L_2 such that $f(S^2x) = -f(x)$, then $f=0$. Since the absolute value of f is a constant, it follows that $f^2 \in L_2$, and, since f^2 is invariant under S^2, it follows that f^2 is equal almost everywhere to a constant.

Write $g(x) = f(Sx)$. Since $g(S^2x) = -g(x)$, the proper value theorem implies that $g(x) = cf(x)$ for some constant c, i.e., that $f(Sx) = cf(x)$. It follows that $f(S^2x) = c^2f(x)$ and hence that $c^2 = -1$. Since $\bar{f}(Sx) = \overline{cf}(x)$, there is no loss of generality in assuming that $c = i$. The fact that proper vectors corresponding to distinct proper values are orthogonal implies that f and $\bar{f}$ are orthogonal, i.e., that $\int f^2(x)dx = 0$. This, combined with the fact that f^2 is a constant, completes the proof.

PROOF OF SUFFICIENCY. Let C be the spectrum of T. Let C^* be a maximal subgroup of the circle such that C^* includes C and such that C^* does not contain -1. I assert that the operation of squaring is an automorphism of C^*. The fact that it is a homomorphism is clear; the fact that its kernel is trivial is implied by its not containing -1. It remains to prove that the square mapping on C^* is onto. If not, then C^* has an element, say a, that has no square root in C^*. Let b be a square root of a in the circle and let D be the group generated by C^* and b. The maximality of C^* implies that $-1 \in D$; this means that $cb^n = -1$ for some element c of C^* and some integer n. Since b^2 is in C^* and -1 is not, the exponent n must be odd, say $n = 2k+1$. Since $cb^{2k+1} = ca^kb$, I conclude that $b = -1/ca^k$ and hence (squaring) that $a = (1/ca^k)^2$. This contradicts the assumption that a has no square root in C^* and proves therefore that squaring is indeed an automorphism of C^*. Let u be the inverse of this automorphism. Regarded as a function from C into the circle, u is a character of C with the property that $(u(c))^2 = c$ for each c in C. If X is the character group of C and if z is the special element of X defined by $z(c) = c$ for all c, then we already know (from the representation theorem) that T is conjugate to the rotation S that sends each x in X onto zx. The rotation that maps x onto ux is a square root of S; the proof of the theorem is complete.

The sufficiency of the condition is true even if T is not ergodic (Amer. J., 1942, p. 159), but who cares?

AUTOMORPHISMS OF COMPACT GROUPS

One of the outstanding problems of ergodic theory is to discover to what extent the conclusions and the implications of the discrete spectrum theorem remain true for transformations with not necessarily discrete spectrum. All that is available at the moment is a meager supply of examples.

In order to study the best known example of a transformation with continuous spectrum, I begin with a comment on complete orthonormal sets in L_2. If $\{f_i\}$ is such a set for a measure space X and $\{g_j\}$ is such a set for Y, and if h_{ij} is defined on the Cartesian product Z of X and Y by $h_{ij}(x, y) = f_i(x)g_j(y)$, then $\{h_{ij}\}$ is a complete orthonormal set for Z. The result extends immediately to the Cartesian product of any finite number of factors. There is even a sense in which the result extends to infinitely many factors. To obtain a basis of L_2 over an infinite product of normalized measure spaces, take a basis (containing the constant function 1) for the L_2 over each of the given spaces and form all products consisting of one factor from each of a finite number of these bases.

If X_0 is the space consisting of the numbers -1 and $+1$ (each carrying the measure $\frac{1}{2}$), then one basis for L_2 consists of the constant function 1 together with the identity function x_0 defined by $x_0(x) = x$. From this fact and from the preceding comments about bases in product spaces it is easy to construct a basis in, say, the bilateral sequence space X of -1's and $+1$'s. (Hitherto we have considered sequences of 0's and 1's. The difference is merely a matter of notational convenience.) If x_n denotes, for every integer n, the n-th coordinate function on X, then a basis for L_2 over X consists of all finite products of the x_n's; the constant function 1 can be forced into this scheme by assigning it as the value of the empty product.

Suppose now that T is the bilateral coordinate shift on X and that U is the induced unitary operator. Call two members of the finite-product basis just described U-equivalent if some integer power of U carries one onto the other. The function 1 constitutes its own U-equivalence class ; the remaining basic functions split into countably many U-equivalence classes each one of which is infinite. Each such U-equivalence class is in an obvious one-to-one correspondence with the set of all integers ; the action of U on the class is to replace the element corresponding to n by the element corresponding to $n+1$. Hence in Hilbert space terms (i.e., to within spectral, or unitary, equivalence) U can be described as follows : there exists a basis consisting of a vector f_0 and of an infinite matrix of vectors $f(i,j)$, where $i=1, 2, 3,\cdots$ and $j=0, \pm 1, \pm 2,\cdots$, such that $Uf_0=f_0$ and such that $Uf(i,j)=f(i,j+1)$ for all i and j. Situations such as this occur often enough to deserve a short name. If a unitary operator U possesses a basis that is like the one just described in all respects except in the cardinality of the index set $\{i\}$, and if that cardinality is n (finite or infinite), I shall say that U is of type n. In this language, the bilateral coordinate shift is of type aleph-null.

If X_0 is an arbitrary normalized measure space, there is no difficulty in forming the space X of all bilateral infinite sequences $\{x_n\}$ with x_n in X_0, $n=0, \pm 1, \pm 2,\cdots$, and defining the Cartesian product measure in X. The bilateral coordinate shift transformation can then also be defined on X ; the techniques used in showing that the two-point shift is ergodic, and in fact strongly mixing, serve to establish the same results for this generalized shift. If X_0 is not too large (precisely speaking : if X_0 is separable, or, equivalently, if the L_2 space over X_0 is separable), then the spectral type of the generalized shift is aleph-null. (The situation is not very different even for non-separable X_0's ; the only thing that changes is the cardinal number associated with the type.) In particular, the three-point shift and the two-point shift are of the same type. In more usual terminology, these two transformations are equivalent. Whether or not they are also conjugate is one of the most exasperating unsolved problems of ergodic theory.

If $X_0=\{-1, +1\}$, then X_0 is a multiplicative abelian group. Endow X_0 with the discrete topology, thus converting it into a compact abelian group, and form the topological Cartesian product X of countably infinitely many copies of X_0. The compact abelian group X is identifiable with the bilateral two-point sequence space. From the group-theoretic point of view the bilateral coordinate shift has an important property: it is a continuous automorphism of X. The generalized shifts can also be identified in this way with continuous automorphisms of certain compact abelian groups. The result that, in the presence of a suitable countability assumption, they are of type aleph-null is a special case of a theorem to be proved in a little while.

Suppose that X is a compact abelian group and let C be its character group. We already know that if T is a continuous automorphism of X, then T is a measure-preserving transformation on X (with respect to Haar measure). Denote by U the unitary operator induced by T. It is easy to verify that if $f \in C$, then $Uf \in C$, and, moreover, the mapping U restricted to C is an automorphism of the group C. In particular, if $f \in C$, then it makes sense to speak of the orbit of f under U; this means, of course, the set of all characters of the form $U^n f$, where n is an integer. If f is the principal character (i.e., the constant function 1), then the orbit of f consists of f alone. If U has no other finite orbits, I shall say, for the sake of brevity, that U has no finite orbits.

AUTOMORPHISM THEOREM. *If a continuous automorphism T of a compact abelian group X is ergodic, then the induced automorphism U on the character group C has no finite orbits. If U has no finite orbits, then T is of the standard type n for some cardinal number n and, therefore, T is strongly mixing.*

PROOF. Suppose that the orbit under U of an f in C ($f \neq 1$) is finite, and let n be the least positive integer such that $U^n f = f$. It follows that the orbit of f consists of f, $Uf, \cdots, U^{n-1}f$; if g is the sum of those functions, then $Ug = g$. The orthogonality, and the consequent linear independence, of distinct characters shows that g is not a constant, and therefore T is not ergodic.

Suppose next that U has no finite orbits. The fact that the characters of X constitute a basis for L_2 implies that T is of type n for some n; all that remains to be shown is that the transformations of that type are necessarily mixing. This is a purely Hilbert space lemma. Considering separately each row of the matrix of basis vectors $f(i, j)$ and recalling the functional form of the definition of strong mixing, I reduce the problem to this one : if $\{f_j\}$ is a basis of a Hilbert space, $j=0, \pm1, \pm2, \cdots$, and if U is the unitary operator such that $Uf_j=f_{j+1}$, then U^n tends weakly to 0, i.e., (U^nf, g) tends to 0 for each f and g. It is sufficient to prove this in case both f and g are basis vectors ; the general case follows by an obvious continuity and linearity argument. The special case, however, is trivial : if $f=f_j$ and $g=f_k$, then $(U^nf, g)=(f_{j+n}, f_k)$, and therefore $(U^nf, g)=0$ for all sufficiently large n.

It is an immediate corollary of this theorem that for continuous automorphisms of compact abelian groups ergodicity is the same as mixing. The theorem has been generalized to non-abelian groups ; see Kaplansky, Can. J., 1949, p. 111.

A stronger form of the theorem is true ; by a purely group-theoretic argument I proved once that the cardinal number n is necessarily infinite (Bull. A.M.S., 1943, p. 621). (In fact, the existence of a measure-preserving transformation of finite type is an open question.) In many special cases the fact that n is infinite is obvious by inspection ; the general case goes as follows.

INFINITE MULTIPLICITY THEOREM. *If an automorphism U of an abelian group C has only infinite orbits (other than the trivial orbit $\{1\}$), then it has infinitely many orbits (provided that $C \neq \{1\}$).*

PROOF. Assume that there are only finitely many orbits. Case I : C is finitely generated. It follows from the basis theorem that C has only finitely many elements of finite order and also that C has only finitely many elements of any given height. (The height of an element f is the supremum of the positive integers n such that $f=g^n$ for some g in C.) Since every element in an orbit has the same order, it follows that C is torsion-free. Since, similarly, every element in an orbit has the same height, and since, for any $f\ (\neq1)$, infinitely many terms of the sequence $\{f^n\}$ must lie in the same orbit, it follows that some orbit contains elements of arbitrarily large heights—a contradiction. Case II : C is arbitrary. It is sufficient to prove that C has a finitely generated non-trivial subgroup that is invariant under U and U^{-1}. If $f \in C$, $f \neq 1$, then there must exist two formally different terms of the sequence $\left\{ \prod_{j=0}^{n} U^jf \right\}$ that lie in the same orbit, say $\prod_{j=0}^{n} U^jf$

$= U^k \left(\prod_{j=0}^{m} U^j f \right) = \prod_{j=0}^{m} U^{j+k} f.$ Cancelling formally equal terms, I obtain
a relation of the form $(U^p f^{\pm 1}) \cdots (U^q f^{\pm 1}) = 1$ with $p < \cdots < q$.   It follows
that the subgroup generated by $U^p f$, $U^{p+1} f$, $\cdots$, $U^q f$ is invariant under U
and U^{-1}.   (This elegant proof is quite a bit simpler than my own original
one; it is due to Rohlin, Izvestya, 1941, p. 329.)

As an important special case of the automorphism theorem, consider
the torus in the role of X.   We already know that every automorphism T
of X is induced by a unimodular matrix $\begin{pmatrix} a & b \\ c & d \end{pmatrix}$ via the equation $T(u, v)$
$=(u^a v^b, u^c v^d)$.   Every character of X is of the form $f(u, v) = u^n v^m$, where
n and m are integers.   It follows that $Uf(u, v) = f(T(u, v)) = u^{an+cm} v^{bn+dm}$,
so that U acts on the character group C of X (i.e., on the lattice points in
the plane) as the transposed matrix $M = \begin{pmatrix} a & c \\ b & d \end{pmatrix}$.   It follows on purely algebraic grounds that U has no finite orbits if and only if no root of unity
is a proper value of M.   (Proof.   If the orbit of (n, m) under M is finite,
say $M^k(n, m) = (n, m)$, let q be a k-th root of unity and observe that
$q^{k-1}(n, m) + q^{k-2} M(n, m) + \cdots + M^{k-1}(n, m)$ is a complex proper vector of M
with proper value q.   If, conversely, M has q as a proper value, then $M^k - 1$
is a singular linear transformation on a rational vector space; it follows
that there exists a non-zero lattice point (n, m) whose orbit under M is
finite.)   These remarks make it possible to write down any number of mixing transformations on the torus.   One such is defined by $T(u, v) = (uv, u)$;
in real terms, $T(x, y) = (x + y, x) \pmod 1$.

Another example is obtained by taking X to be the (multiplicatively
written) character group of the (additive) group of rational numbers (or,
for a slightly different example, dyadic rational numbers); on this group
the operation of squaring is a mixing automorphism.

The various examples so obtained do not solve the problem of equivalence versus conjugacy for transformations with continuous spectrum.
They (i.e., the examples) are all equivalent (if, that is, the groups are sub-
jected to a countability restriction), and they are not likely to be conjugate.
Since, however, they have not been proved to belong to different conjugate classes, the discussion leaves the problem open.

It is worth while to note that although no tool such as the representation theorem for transformations with discrete spectrum is available for group automorphisms, some of the consequences of that theorem are valid for an interesting class of such automorphisms, namely for shifts. Consider the bilateral sequence space based on a normalized measure space X and let T be the bilateral shift on the sequence space. If the transformations P and Q are defined by $(Px)_n = x_{-n}$ and $(Qx)_n = x_{1-n}$, respectively, then $PQ = T$, so that every shift is the product of two involutions (and, consequently, every shift is similar to its own inverse).

Write T_X for the bilateral shift based on X. It is clear that if X and Y are measure-theoretically isomorphic spaces, i.e., if there exists an invertible measure-preserving transformation mapping X onto Y, then T_X and T_Y are similar transformations. Observe next that the Cartesian product of two shifts is in a natural way similar to the shift based on the Cartesian product space, i.e., that $T_X \times T_Y$ is similar to $T_{X \times Y}$. Observe finally that the square of a shift is similar to its Cartesian square, i.e., that T_X^2 is similar to $T_X \times T_X$. It follows from these remarks that if, for instance, X consists of 9 points, each with measure $1/9$, then T_X has a square root that is similar to the shift based on a space with 3 points, each with measure $1/3$. It follows also, since the unit interval is measure-theoretically isomorphic to the unit square, that if X is the unit interval, then T_X has a square root that is similar to T_X itself.

GENERALIZED PROPER VALUES

Suppose that T is an invertible measure-preserving transformation on a normalized measure space X. I propose to study certain classes of functions associated with T and a certain method of constructing new classes out of old. The classes of interest will all consist of measurable functions of constant absolute value 1. If G is such a class, let G' be the class of all those functions f (measurable and of constant absolute value 1) for which there exists a function g in G such that $f(Tx) = g(x)f(x)$ almost everywhere. In other words, the elements of G' may be viewed as proper vectors belonging to generalized proper values; the proper values, instead of being necessarily constants, are elements of G. It is obvious, but useful to note, that if $H \subset G$, then $H' \subset G'$.

Let G_1 be the set of all constant functions (of absolute value 1) and define G_n inductively by $G_{n+1} = G'_n$, for $n = 1, 2, 3, \cdots$. The elements of G_2, in particular, are proper functions; if we assume (as we do from here on) that T is ergodic, then every proper function of T is in G_2, except for a multiplicative constant factor. If f is a constant, then $f(Tx) = 1 \cdot f(x)$; since $1 \in G_1$, it follows that $f \in G_2$, so that $G_1 \subset G_2$. From this in turn it follows inductively that the sequence $\{G_n\}$ is increasing. I mention in passing that each G_n is a multiplicative group and that each G_n is invariant under T (i.e., $Uf \in G_n$ if and only if $f \in G_n$, where U, of course, is the unitary operator induced by T); the proofs of these facts are also easy inductions.

If it ever happens that $G_n = G_{n+1}$, then $G_n = G_{n+k}$ for all k. I shall denote by $n(T)$ the least positive integer for which this does happen; I do not rule out the possibility that $n(T)$ is infinite. In any case the function n is obviously a conjugacy invariant of T; if, in other words, S and T are conjugate, then $n(S) = n(T)$. I shall prove the existence of two equivalent but non-conjugate transformations by exhibiting two equivalent transformations S and T such that $n(S) \neq n(T)$. Observe that $n(T) = 1$ is characteristic of transformations with continuous spectrum.

To illustrate the technique of computing $n(T)$, suppose that X is the circle and that R is an ergodic rotation, $Rx = cx$. The set G_2 consists of all proper functions of R, i.e., of all the functions g_m defined by $g_m(x) = x^m$, $m = 0, \pm 1, \pm 2, \cdots$, and their constant multiples. Suppose that $f \in G_3$, so that $Uf = bg_mf$ for some integer m and for some constant b of absolute value 1. Expand f in a Fourier series, $f = \sum a_n g_n$; since $Uf = \sum a_n c^n g_n$ and $bg_mf = \sum a_{n-m} bg_n$, it follows that $a_{n-m}b = a_n c^n$. Since this implies that $|a_{n-m}| = |a_n|$ for all n, I conclude that $m = 0$; otherwise each a_n would have to be 0, contradicting the assumption that $|f| = 1$. (Recall that $\sum |a_n|^2$ converges.) If, however, $m = 0$, then f is an ordinary proper vector of U, so that $f \in G_2$. This proves that $G_3 = G_2$ and hence that $n(R) = 2$.

The counterexamples S and T mentioned earlier are defined as follows. The space X is the torus. Let c be a complex number of absolute value 1 that is not a root of unity, and let Q be a mixing transformation of type aleph-null on the circle. I write $S(x, y) = (cx, xy)$ and $T(x, y) = (cx, Qy)$. I shall show that S and T are equivalent and that $n(S) = 3$ and $n(T) = 2$. The symbols U and V will denote the unitary operators induced by S and T respectively.

Write $g_{n,m}(x, y) = x^n y^m$, $n, m = 0, \pm 1, \pm 2, \cdots$; the functions $g_{n,m}$ constitute a complete orthonormal set in L_2. Since $Ug_{n,m} = c^n g_{n+m,m}$, the function $g_{n,0}$ is a proper function with proper value c^n, and the functions $g_{n,m}$ with $m \neq 0$ get permuted among themselves and multiplied by certain constant factors at the same time. I want to get rid of those constant factors. In other words, I want to write $f_{n,m} = a_{n,m} g_{n,m}$, where $a_{n,m}$ is a constant of absolute value 1 whenever $m \neq 0$, and to do this in such a way that $Uf_{n,m} = f_{n+m,m}$ when $m \neq 0$. This requirement imposes certain conditions on the constants $a_{n,m}$; a detailed examination of the situation shows that the conditions can be satisfied by an appropriate inductive procedure. They can also be satisfied by an explicit formula: choose b_m so that $b_m^{8m} = c$ and write $a_{n,m} = b_m^{(2n-m)^2}$. After an obvious change of notation, the result obtained can be described as follows. There exists a complete orthonormal set consisting of a sequence $\{h_n\}$ and a double sequence $\{h_{i,j}\}$ such that h_n is a proper vector with proper value c^n and such that $h_{i,j}$ goes onto $h_{i+1,j}$, where $i = 0, \pm 1, \pm 2, \cdots$, and $j = 1, 2, 3, \cdots$. This is a complete spectral characterization of S.

I turn now to the calculation of $n(S)$. Since every function in G_2 is a constant multiple of some x^m, it follows that if $f \in G_3$, then $f(cx, xy) = bx^m f(x, y)$ for some integer m and some constant b with $|b| = 1$. For each fixed x the function $f(x, y)$ can be expanded in a Fourier y-series whose coefficients are measurable functions of x. If $f(x, y) = \sum f_n(x)y^n$, then $f(cx, xy) = \sum f_n(cx)x^n y^n$ and $bx^m f(x, y) = \sum b f_n(x)x^m y^n$, so that $f_n(cx) = bx^{m-n}f_n(x)$. From our evaluation of $n(R)$ for an ergodic rotation R it follows that the only non-zero possibility is $n = m$ and that the function f_m must itself be a proper function of R. It follows that $f(x, y)$ must be a constant multiple of $x^k y^m$ for some k. It is trivial to verify that, conversely, if f is of this form, then $f \in G_3$. The result implies, in particular, that $n(S) \geqq 3$.

If $f \in G_4$, then $f(cx, xy) = bx^k y^m f(x, y)$. Expand as before. If $f(x, y) = \sum f_n(x)y^n$, then $f(cx, xy) = \sum f_n(cx)x^n y^n$ and $bx^k y^m f(x, y) = \sum b f_{n-m}(x)x^k y^n$, so that $f_n(cx)x^n = bf_{n-m}(x)x^k$. Forming the norms of both sides, I conclude that $\|f_n\| = \|f_{n-m}\|$ for all n. Since $\sum \|f_n\|^2$ converges, this is impossible unless $m = 0$. If, moreover, $m = 0$, then $f_n(cx) = bx^{k-n}f_n(x)$; it follows as before that the only non-zero possibility is $n = k$, and that $f_k(x)$ must be a constant multiple of some power of x. This in turn implies that $f \in G_3$; we have proved that $G_4 = G_3$ and hence that $n(S) = 3$.

The situation for T is similar. Let W be the unitary operator induced by the auxiliary mixing transformation Q of type aleph-null. Let $\{g_{n,m}\}$ be an orthonormal set for the circle ($n = 0, \pm 1, \pm 2, \cdots, m = 1, 2, 3, \cdots$) such that together with the constant function 1 it forms a basis and such that $Wg_{n,m} = g_{n+1,m}$. Write $h_n(x, y) = x^n$, and $p_{k,n,m}(x, y) = x^k g_{n,m}(y)$. Clearly $Vh_n = c^n h_n$ and $Vp_{k,n,m} = c^k p_{k,n+1,m}$. Get rid of the factor c^k, as once before in the case of U. An efficient way of doing this is to write $q_{k,n,m} = c^{mk}p_{k,n,m}$; it follows that $Vq_{k,n,m} = q_{k,n+1,m}$. The set of all pairs (k, m) is countably infinite and, therefore, may be put into one-to-one correspondence with the positive integers j. If $h_{i,j}$ is used to denote $q_{k,i,m}$ whenever (k, m) corresponds to j, then $Vh_{i,j} = h_{i+1,j}$; this completes the proof that U and V are equivalent.

To prove that $n(T) = 2$, suppose that f belongs to the class G_3 for T, i.e., that $f(cx, Qy) = bx^k f(x, y)$ for some integer k and some constant b of absolute value 1. Expand $f(x, y)$ in a y-series in terms of the $g_{n, m}$'s. If $f(x, y) = f_0(x) + \sum f_{n, m}(x) g_{n, m}(y)$, then $f(cx, Qy) = f_0(cx) + \sum f_{n-1, m}(cx) g_{n, m}(y)$ and $bx^k f(x, y) = bx^k f_0(x) + \sum bx^k f_{n, m}(x) g_{n, m}(y)$. It follows that $f_0(cx) = bx^k f_0(x)$ and therefore that f_0 is a constant multiple of some power of x. It follows also that $f_{n-1, m}(cx) = bx^k f_{n, m}(x)$. Taking norms, I conclude, as once before, that $f_{n, m} = 0$ for all n and m. This implies that $f(x, y) = f_0(x)$ and hence that $f \in G^2$. The proof is complete.

The result so obtained is very special. We know now, to be sure, that equivalence does not imply conjugacy. Since, however, the equivalent transformations S and T have mixed spectrum (neither discrete nor continuous), we do not know whether equivalence implies conjugacy for transformations with continuous spectrum, nor, in particular, whether any two mixing transformations of type aleph-null are conjugate. The problem is still open. The mixed-spectrum example has some value, however; the method (i.e., the consideration of generalized proper values and generalized proper functions) is of interest apart from the special result that it was used to derive. I keep having the feeling, for instance, that generalized proper values could yield some new information in spectral theory.

The transformation S is a special case of what Anzai called a skew product (Osaka J., 1951, p. 83); using such skew products, Anzai constructed a measure-preserving transformation that is not conjugate to its inverse.

WEAK TOPOLOGY

What sense does it make to speak of a sequence of measure-preserving transformations converging to a measure-preserving transformation? In other words, what interesting topologies can be introduced into the set of all measure-preserving transformations on a measure space? There are several, with various virtues; I proceed to study one of the most fruitful ones.

Some preliminary comments are necessary. First of all, the discussion will be limited to invertible measure-preserving transformations, so that the set to be topologized is, in fact, a group. Secondly, two measure-preserving transformations will be identified, as usual, if they differ on a set of measure zero only; this makes it reasonable to concentrate attention not on the group of invertible measure-preserving transformations on a measure space X but on the group of automorphisms of the measure algebra B associated with X. Lastly (and leastly) the underlying measure space X will be assumed to be the unit interval. This last assumption is not as special as it looks. The measure algebra of the interval is normalized, non-atomic, and separable. It is known that every two measure algebras satisfying these three conditions are isomorphic. The fact that X is the unit interval will be used in two ways. The main use will be based on the fact that B is normalized, non-atomic, and separable; all considerations following from these properties apply to most of the other measure spaces I have ever mentioned (e.g., to every non-discrete compact group with a countable base, and, in particular, to the torus and to the sequence space). Incidental use will be made of some of the order properties of the interval (e.g., subintervals will be mentioned). Such considerations do not generalize verbatim, of course, but, in order to generalize them, all that is needed is to replace subintervals by those elements of the appropriate Boolean algebra that correspond to subintervals under the appropriate isomorphism.

61

A final preliminary comment concerns the relation of the topology to be studied to the various known topologies for operators on a Hilbert space. The strong topology for operators is the one for which $\{A_j\}$ converges to A if and only if $\|A_j f - A f\|$ converges to 0 for each f; the weak topology demands that $(A_j f, g)$ converge to $(A f, g)$ for each f and g. (The index set $\{j\}$ here is an arbitrary directed set.) Since automorphisms of B are (induce) unitary operators on L_2, the group of automorphisms inherits every topology that the group of unitary operators possesses. The strong and the weak topologies restricted to unitary operators happen to coincide. (Proof. Strong convergence always implies weak convergence. If the U_j's are unitary operators converging weakly to a unitary operator U, multiply out the square norm $\|U_j f - U f\|^2$ and observe that each of the four terms tends, except for sign, to (f, f).) It follows that the specializations of these topologies to the group G of automorphisms coincide; the topology I am about to describe (to be called the weak topology) is what they collapse to.

The weak topology for the group G of automorphisms is the one for which $\{T_j\}$ converges to T if and only if $\{T_j E\}$ converges to TE for each measurable set E (or, rather, for each element E of the measure algebra B); more explicitly, it is required that $m(T_j E + TE)$ tend to 0. (The plus sign refers to Boolean addition, i.e., to symmetric difference.) Still more explicitly, a subbasis for the open sets consists of all sets of the form

$$N(S) = N(S; E, \varepsilon) = \{T: m(SE + TE) < \varepsilon\}.$$

The first result concerning the weak topology is that with respect to that topology the group G becomes a topological group satisfying the first countability axiom. To prove that G is a T_0-space, suppose that $S \neq I$ (the symbol I denotes the identity automorphism) and let E be a set (i.e., an element of B) such that $SE \neq E$. If $\varepsilon = m(SE + E)$, then the neighborhood $N(I; E, \varepsilon)$ does not contain S. The continuity of the inverse operation follows from the fact that $T \in N(S; E, \varepsilon)$ if and only if $T^{-1} \in N(S^{-1}; T^{-1}E, \varepsilon)$. To prove the continuity of multiplication, suppose that $\{S_j\}$ and $\{T_j\}$ converge to S and to T respectively. Since $T_j E$ is close to TE when j is large and $S_j TE$ is close to STE when j is large, and since the measure-preserving character of S_j implies that $S_j T_j E$ is just as close to $S_j TE$ as $T_j E$ is to TE, it follows that $S_j T_j E$ is close to STE when j is large. These remarks prove that G is a topological group. To prove the first countability axiom it is therefore sufficient to prove the existence of a countable base at I. Select, for this purpose, a countable dense set $\{E_k\}$ in B; the neighborhoods $N(I; E_k, 1/h)$ constitute a subbasis at I. Suppose, indeed, that $N(I; E, \varepsilon)$ is an arbitrary subbasic neighborhood of I. If h and k are chosen so that both $m(E+E_k)$ and $1/h$ are less than $\varepsilon/3$, then $N(I; E_k, 1/h) \subset N(I; E, \varepsilon)$.

Associated with G (as with every topological group) there are some natural concepts of uniformity. Let us say that the left uniformity is the one according to which $\{T_n\}$ is a Cauchy sequence if and only if $T_n^{-1}T_m$ is near to I whenever n and m are large; the right uniformity is defined similarly in terms of $T_nT_m^{-1}$. The group G does not happen to be complete with respect to either of these uniformities. The pertinent examples are most easily given in the unilateral sequence space. Let T_n be the transformation induced by a cyclic permutation of the first n indices, i.e., if $T_nx=y$, then $y_k=x_{k+1}$ whenever $0 \leqq k < n$, $y_n=x_0$, and $y_k=x_k$ whenever $k>n$. It is clear that each T_n is an invertible measure-preserving transformation. I assert that T_nE converges, for each E, to TE, where T is the unilateral shift. For finite-dimensional sets this is easy (T_nE is, in fact, equal to TE for all large n); for arbitrary measurable sets it follows by approximation. It follows that $\{T_n\}$ is a left Cauchy sequence; since, however, the unilateral shift is not invertible, the sequence has no limit in G. Consideration of the sequence $\{T_n^{-1}\}$ shows that G is not right complete either.

The correct concept of uniformity in G is the ambidextrous one, i.e., the one according to which $\{T_n\}$ is a Cauchy sequence if and only if both $T_n^{-1}T_m$ and $T_nT_m^{-1}$ are near to I whenever n and m are large. With respect to this uniformity G is complete. In view of the first countability axiom it is sufficient to prove sequential completeness. If $\{T_n\}$ is an ambidextrous Cauchy sequence in G, then both $\{T_nE\}$ and $\{T_n^{-1}E\}$ are Cauchy sequences in B; it follows that $\{T_nE\}$ and $\{T_n^{-1}E\}$ converge to, say, TE and SE, respectively, for each E in B. (The completeness of B is tantamount to the Riesz-Fischer theorem.) Since complement, union, and measure are continuous functions on B, the mappings T and S on B are measure-preserving and preserve all finite Boolean operations;

from this it follows that they preserve the countable operations also. (In general a Boolean homomorphism need not be a sigma-homomorphism; the measure is what makes things work here.) The mappings T and S are therefore endomorphisms of B. Since it is easily verified that $TSE = STE = E$ for all E, it follows that they are, in fact, automorphisms, and that $S = T^{-1}$.

The concept of completeness is more familiar for metric spaces than for more general uniform spaces (e.g., topological groups). Since it is known that a uniform space satisfying the first countability axiom is metrizable, the properties of G could, in principle, have been discussed in terms of a suitable metric. It is easy to exhibit such a metric. Let $\{E_n\}$ be a countable dense set in B and define the distance between S and T to be $\sum(1/2^n)(m(SE_n + TE_n) + m(S^{-1}E_n + T^{-1}E_n))$; the assertion is that this metric induces the weak topology and the ambidextrous uniform structure on G. (It is neither left nor right invariant.) Since, however, this rather artificial construction does not seem to throw any light on the structure of G, I see no point in studying it further. The situation would be quite different if an invariant metric could be found (i.e., a metric invariant under both left and right multiplications in G). The general theory of topological groups guarantees the existence of a one-sided invariant metric; a two-sided one, however, does not exist. For a slightly more detailed account of some of these side issues, and for references to the literature, see my first paper on this subject (Trans. A.M.S., 1944, p. 11).

WEAK APPROXIMATION

I shall call an interval $(k/2^n, (k+1)/2^n)$ a dyadic interval of rank n $(n = 0, 1, 2, \cdots; k = 0, 1, \cdots, 2^n - 1)$ and a union of such intervals (n fixed) a dyadic set of rank n. By a permutation (more precisely, a dyadic permutation of rank n) I shall mean an invertible measure-preserving transformation (or, rather, the corresponding automorphism) that maps each dyadic interval of rank n onto a dyadic interval of rank n by an ordinary translation. A cyclic permutation of rank n is a permutation that acts as a cyclic permutation on the dyadic intervals of rank n. (Caution: this means that there is only one cycle, not that each dyadic interval of rank n gets mapped onto the next one in the natural order.) A dyadic sub-basic neighborhood (of an automorphism S in G) is a set of the form $\{T: m(SD + TD) < \varepsilon\}$, where D is a dyadic set. The dyadic neighborhoods constitute a subbasis for the topology of G; finite intersections of them constitute a basis. The fundamental theorem concerning these concepts asserts essentially that the permutations are dense in G; the following more quantitative formulation is sometimes useful.

WEAK APPROXIMATION THEOREM. *Every dyadic neighborhood contains cyclic permutations of arbitrarily high ranks.*

PROOF. I prove first that if P is a permutation and if $N = \{T: m(PD_i + TD_i) < \varepsilon, i = 1, \cdots, n\}$ is a dyadic neighborhood of P, then N contains cyclic permutations of arbitrarily high ranks; after that I shall complete the proof by showing that the permutations are dense in G. Each PD_i is a dyadic set. It follows that there exists an integer j such that each of these sets is dyadic of rank j; the integer j can be selected as large as desired. Choose k greater than j and so large that $1/2^k < \varepsilon/2^{j+1}$. Each dyadic interval of rank j is a union of 2^{k-j} dyadic intervals of rank k. I shall define a cyclic permutation Q of rank k so that $Q \in N$.

Begin with a dyadic interval E of rank j. If P does not leave E fixed, let Q map the first (natural order) subinterval of rank k in E onto the first

subinterval of rank k in PE. If P does not return PE to E, let Q map
the first subinterval of rank k in PE onto the first subinterval of rank k
in P^2E. Continue in this way till the process reaches the last term in
the P-cycle of E, say $P^{q-1}E$. The permutation P returns $P^{q-1}E$ to E;
let Q map the first subinterval of rank k in $P^{q-1}E$ onto the second subinterval of rank k in E. Now go through the cycle again with " second "
replacing " first " everywhere. When the end is reached, let Q send
the second subinterval of rank k in $P^{q-1}E$ onto the third subinterval of
rank k in E. Repeat again and again till the last subinterval of rank k in
$P^{q-1}E$ is reached. Let F be a dyadic interval of rank j disjoint from the P-
cycle of E (i.e., distinct from $E, \cdots, P^{q-1}E$) and let Q map the last subinterval of rank k in $P^{q-1}E$ onto the first subinterval of rank k in F. Then
repeat with F what was just done with E, and proceed in this way through
the entire cycle decomposition of P. The permutation Q is to send the
last subinterval of rank k in the last term of the last cycle of P onto the
first subinterval of rank k in the originally chosen interval E.

It is clear that Q is a cyclic permutation of rank k. The construction
of Q implies that if E is any dyadic interval of rank j (not necessarily the
particular E used above), then $m(PE+QE)<2/2^k$. Since a dyadic set
of rank j is the union of at most 2^j dyadic intervals of rank j, it follows
that if E is an arbitrary dyadic set of rank j, then
$$m(PE+QE)<2^j \cdot (2/2^k)=2^{j+1}/2^k<\varepsilon.$$
This implies that $Q \in N$; it remains to prove that permutations are dense
in G.

I must show that if T is an arbitrary transformation and if $N=\{S:$
$m(SD_i+TD_i)<\varepsilon, i=1,\cdots, n\}$ is a dyadic neighborhood of T, then N contains a permutation. I may assume that the dyadic sets D_i consist of all
the dyadic intervals of a certain rank (so that n is a power of 2); a neighborhood defined by such intervals and a sufficiently small ε is always included
in any dyadic neighborhood. The idea of the proof is to approximate the
TD_i's by dyadic sets and then to permute those dyadic sets as necessary.

The class B_0 of dyadic sets is dense in the metric space of sets (or,
rather, in the measure algebra), it is closed under the finite Boolean operations, and the measure of each set in it is a dyadic rational number. If,
moreover, $E \in B_0$ and if r is any dyadic rational number between 0 and

$m(E)$, then E has subsets that belong to B_0 and have measure r. These are the crucial properties of B_0 that will enable me to prove the pertinent approximation lemma. The reason for listing these properties is that there are classes other than B_0 that have the same properties and that it will be necessary to use; the most notable example of such a class is the class of all sets of the form TE with E in B_0.

Suppose now that $\{E_1, \cdots, E_k\}$ is a partition of X, that δ is a positive number, that $r_1, \cdots, r_k$ are positive dyadic rational numbers whose sum is 1, and that $|m(E_i) - r_i| < \delta$ for $i = 1, \cdots, k$. I assert that there exists a partition $\{F_1, \cdots, F_k\}$ of X such that each F_i is in B_0, such that $m(E_i + F_i) < 2\delta$, and such that $m(F_i) = r_i$, $i = 1, \cdots, k$. To prove this, begin by approximating each E_i very closely by a set F_i in B_0; let the degree of approximation be denoted by γ. The sets F_i, as they stand, are not necessarily a partition of X and do not necessarily have the right measure. Since $m(E_i) - m(F_i) = m(E_i - F_i) - m(F_i - E_i)$, it follows that $m(F_i)$ is within γ of $m(E_i)$ and hence within $\gamma + \delta$ of r_i. I now proceed to change the F_i's; my first object is to make them disjoint. Since $m((E_i \cap E_j) + (F_i \cap F_j)) \leqq m(E_i + F_i) + m(E_j + F_j) < 2\gamma$, and since $m(E_i \cap E_j) = 0$, it follows that the measure of the intersection of each F_i with the union of the others is less than $2k\gamma$. The modified F_i's are the sets obtained by discarding those intersections. It follows that the measure of the new F_i is within $2k\gamma + \gamma + \delta$ of r_i, and it follows also that the new F_i approximates E_i to within $2k\gamma + \gamma$. The F_i's are now disjoint, but their union is not necessarily X, and their measure is still not quite right. One more change is necessary. From each fat F_i (i.e., if $m(F_i) > r_i$) I subtract enough to make its measure equal to r_i, and I throw the discard into the common complement of the F's; then I add enough to each thin F_i from that common complement so as to bring its measure up to r_i and to keep F_i in B_0. Since the new new F_i approximates E_i to within $(2k\gamma + \gamma) + (2k\gamma + \gamma + \delta)$, the proof is completed by the observation that if γ is chosen small enough, then this error term is less than 2δ.

Return now to the neighborhood N. The sets $D_i \cap TD_j$ constitute a partition of X. Let δ be a positive number such that $n^2\delta < \varepsilon/2$. A weak application of the approximation lemma yields a partition $\{E_{ij}\}$ of X into dyadic sets such that $m((D_i \cap TD_j) + E_{ij}) < \delta$. Another application of the same lemma (this time to the T transforms of the dyadic sets) yields a partition $\{F_{ij}\}$ of X into dyadic sets such that $m((D_i \cap TD_j) + TF_{ij}) < \delta$ and such that $m(F_{ij}) = m(E_{ij})$. Since T is measure-preserving, it follows that $m((T^{-1}D_i \cap D_j) + F_{ij}) < \delta$.

Let k be an integer such that all the D_i's, E_{ij}'s, and F_{ij}'s have rank k. Let P be a permutation of rank k that maps F_{ij} onto E_{ij}. Since D_j is the union (over i) of $T^{-1}D_i \cap D_j$, it follows that D_j is within $n^2\delta$, and therefore within $\varepsilon/2$, of the union (over i) of F_{ij}. This implies that PD_j is within $\varepsilon/2$ of the union (over i) of E_{ij}. Since this latter union is within $\varepsilon/2$ of the union (over i) of $D_i \cap TD_j$, and hence of TD_j, it follows that PD_j is within ε of TD_j; this completes the proof of the weak approximation theorem.

UNIFORM TOPOLOGY

The study of various topologies and of the relations among them is, despite its current popularity in the theory of topological linear spaces, a pretty dull business. The fact that the weak and the strong operator topologies coincide when they are restricted to measure-preserving transformations is a godsend : one fewer topology to worry about. It turns out, in fact, that measure-preserving transformations (or, rather, automorphisms) admit exactly two useful topologies (instead of the confusing profusion in operator theory). Before continuing with the applications of the weak topology, I shall now introduce the other useful topology for automorphisms ; I shall call it the uniform topology.

If S and T are elements of the group G of automorphisms, write $d(S, T)$ for the supremum (over all E) of $m(SE + TE)$. It is clear that d is a metric and it is easy to prove that d is invariant under all the group operations (i.e., left multiplication, right multiplication, and the formation of inverses). The uniform topology is, by definition, the topology induced by the metric d ; endowed with the uniform topology, G is a topological group.

A useful tool for studying the distance d is another distance d' ; the definition of d' is

$$d'(S, T) = m(\{x : Sx \neq Tx\}).$$

The verification that d' is a metric is straightforward. The triangle inequality comes from the fact that if Sx is different from Tx, then either Sx or Tx must be different from Qx. The metric d' is invariant under all group operations. For left multiplication this is obvious. To prove the invariance of d' under inverse formation, observe that $S\{x: Sx \neq Tx\} = \{x: S^{-1}x \neq T^{-1}x\}$. Invariance under right multiplication follows from left invariance and inverse invariance together.

Now I digress to pick up a couple of auxiliary facts. They have some interest of their own, and they will be used later ; the immediate purpose in studying them is to clarify the relation between d and d'.

To begin with, I need the usual elementary facts about periodicity. If $T^n x = x$ for some positive integer n, the transformation T is said to be periodic at x; the smallest positive n is the period of T at x. The corresponding global concept is also standard: if $T^n = I$ for some integer n, the transformation T is said to be periodic, and the period of T is defined to be the smallest positive n with this property. Clearly a periodic transformation is periodic at every x; the converse is not true. If T is almost nowhere periodic, I shall say that T is antiperiodic.

Every transformation T breaks up in a natural way into periodic and antiperiodic pieces. Specifically, write A_n for the set of those points at which T is periodic of period n, and let A_0 be the complement of the union of all these A_n's; it follows that T is antiperiodic in A_0.

LEMMA 1. *If T is periodic of period n at (almost) every point of X, then there exists a measurable set E of measure $1/n$ such that the sets E, $TE, \cdots, T^{n-1}E$ are pairwise disjoint.*

PROOF. If $n = 1$, there is nothing to prove. If $n > 1$, there must exist a measurable set E_1 such that $m(E_1 + TE_1) > 0$; otherwise T would be periodic of period 1 at almost every point. Since $m(E_1 - TE_1) = m(E_1) - m(E_1 \cap TE_1)$ and $m(TE_1 - E_1) = m(TE_1) - m(E_1 \cap TE_1)$, the measure-preserving character of T implies that $m(E_1 - TE_1) > 0$. If, in other words, $F_1 = E_1 - TE_1$, then F_1 is a measurable set •of positive measure such that F_1 and TF_1 are disjoint. If $n = 2$, I stop here. If $n > 2$, I assert that there must exist a measurable subset E_2 of F_1 such that $m(E_2 + T^2 E_2) > 0$; otherwise T would be periodic of period 2 at almost every point of F_1. Write $F_2 = E_2 - T^2 E_2$; it follows as before that F_2 and $T^2 F_2$ are disjoint. Continuing in this way for $n - 1$ steps, I find a decreasing sequence $F_1, \ldots, F_{n-1}$ of sets of positive measure such that F_j and $T^j F_j$ are disjoint. If $E = F_{n-1}$, then E is disjoint from $T^j E$, $j = 1, \cdots, n-1$, and this implies that the sets E, $TE, \cdots, T^{n-1}E$ are disjoint. The only thing that may be wrong is that $m(E)$ may not be $1/n$. To rectify this, consider the complement of the set $E \cup TE \cup \cdots \cup T^{n-1}E$, and apply the same argument to that set in place of X. (It is true, but irrelevant, that the set is invariant.) Continue in this way by induction, into the transfinite if necessary. Since a disjoint class of sets of positive measure is necessarily countable, the process will terminate at some countable ordinal.

LEMMA 2.   *If T is antiperiodic, then for every positive integer n and for every positive number ε there exists a measurable set E such that the sets E, TE, $\cdots$, $T^{n-1}E$ are pairwise disjoint and such that $m(E \cup TE \cup \cdots \cup T^{n-1}E) > 1 - \varepsilon$.*

PROOF.   Let p be a positive integer such that $1/p < \varepsilon$. The first main step in the proof is to apply the proof of Lemma 1 (with pn in place of n). What that proof does is this: it constructs a measurable set F of positive measure such that F, TF, $\cdots$, $T^{pn-1}F$ are pairwise disjoint, and such that F is maximal in the measure-theoretic sense. (This means that no set that includes F and has larger measure than F has the stated properties.) The maximality of F implies that if F_0 is a measurable subset of $T^{pn-1}F$, of positive measure, then $T^j F_0 \cap F$ must have positive measure for some $j = 1, \cdots, pn$. (Otherwise TF_0 could be united with F, contradicting maximality.) This, in turn, implies that if F_j is the set of those points x in $T^{pn-1}F$ for which $T^j x$ is in F but $T^i x$ is not in F, $1 \leq i < j \leq pn$, then the pairwise disjoint sets F_j almost exhaust $T^{pn-1}F$.

The sets in each column of the array

$$
\begin{array}{lll}
TF_2 & & \\
TF_3 & T^2F_3 & \\
\vdots & \vdots & \\
TF_{pn} & T^2F_{pn} \cdots T^{pn-1}F_{pn}
\end{array}
$$

are disjoint from one another (since they are transforms by the same power of T of disjoint subsets of $T^{pn-1}F$), and two sets in distinct columns are disjoint from each other (since their inverse images under a suitable power of T are included in distinct ones among the sets F, TF, $\cdots$, $T^{pn-1}F$). I assert, moreover, that all these sets (i.e., the sets $T^i F_j$ with $1 \leq i < j \leq pn$) are disjoint from each $T^k F$ $(0 \leq k \leq pn - 1)$ and hence from $F \cup TF \cup \cdots \cup T^{pn-1}F$. Indeed, if $k > i$, then $T^i F_j \cap T^k F$ is included in $T^i(T^{pn-1}F \cap T^{k-i}F)$, which is empty; if, on the other hand, $k \leq i$, then $0 \leq i - k < j$, so that $T^{i-k}F_j \cap F$ is empty (by the definition of F_j), and therefore $T^i F_j \cap T^k F = T^k(T^{i-k}F_j \cap F)$ is empty also.

The preceding considerations imply, in particular, that the sets TF_1, $T^2F_2, \cdots, T^{pn}F_{pn}$ are pairwise disjoint subsets of F; since their measures are equal to those of F_1, $F_2, \cdots, F_{pn}$, respectively, and hence add up to $m(T^{pn-1}F)$, it follows that they almost exhaust F. From this I infer that the set

$$F^* = \bigcup_{k=0}^{pn-1} T^k F \cup \bigcup_{1 \leq i < j \leq pn} T^i F_j$$

is almost invariant under T. The antiperiodicity of T implies now that F^* is almost equal to X, for otherwise the complement of F^* would furnish a way of enlarging the maximal set F.

I am now ready to define the desired set E. Roughly speaking, E consists of every n-th one among the sets $F, TF, \cdots, T^{pn-1}F$ and also of every n-th one in each row of the array $\{T^i F_j\}$, as far as possible. More precisely,

$$E = \bigcup_{k=0}^{p-1} T^{kn} F \cup \bigcup_{i+n-1 < j} T^i F_j.$$

Clearly the sets $E, TE, \cdots, T^{n-1}E$ are pairwise disjoint. What their union does not contain consists of fewer than n sets from each row of the array $\{T^i F_j\}$. This means that, for each j, the part of the j-th row that is not included in $E \cup TE \cup \cdots \cup T^{n-1}E$ has measure less than $n \cdot m(F_j)$, and, consequently,

$$m(E \cup TE \cup \cdots \cup T^{n-1}E) > \sum_{j=1}^{pn} n \cdot m(F_j) = n \cdot m(F).$$

Since the sets $F, TF, \cdots, T^{pn-1}F$ are pairwise disjoint, it follows that $n \cdot m(F) \leq n \cdot \frac{1}{pn} = \frac{1}{p} < \varepsilon$, and the proof is complete. (The idea of this proof was shown to me by D.S. Ornstein.)

COMPARISON THEOREM. $\frac{2}{3} d' \leq d \leq d'$.

PROOF. Since both d and d' are invariant under the group operations, it is sufficient to prove that $\frac{2}{3} d'(I, T) \leq d(I, T) \leq d'(I, T)$ for every measure-preserving transformation T. If $F = \{x : x \neq Tx\}$, then $d'(I, T) = m(F)$; note that F is invariant under T and so also is every subset of $X - F$. If E is an arbitrary measurable set, then

$$m(E + TE) \leq m((E \cap F) + T(E \cap F)) + m((E - F) + T(E - F))$$
$$= m((E \cap F) + (TE \cap F)) \leq m(F);$$

it follows that $d(I, T) \leq d'(I, T)$.

To prove the lower inequality, let $\{A_0, A_1, A_2, \cdots\}$ be the partition of X into its antiperiodic and periodic pieces. Apply Lemma 2 (with A_0 in place of X, and with $n = 2$ and $\varepsilon \leq 1/3$) to find a measurable subset E_0

of A_0 such that E_0 is disjoint from TE_0 and such that $m(E_0) \geqq \frac{1}{3} m(A_0)$. Apply Lemma 1 to A_n to find a measurable subset E_n of A_n such that the sets $E_n, TE_n, \cdots, T^{n-1}E_n$ are pairwise disjoint and such that $m(E_n) = \frac{1}{n} m(A_n)$.

Define a new sequence of sets F_n as follows. If $n = 0$, then $F_n = E_n$. If $n \geqq 2$, then F_n is the union of E_n and of the images of E_n under even powers of T up to but (in case n is odd) not including $T^{n-1}E_n$. If $n \neq 1$, then F_n is disjoint from TF_n. If, moreover, $n = 2, 4, 6, \cdots$, then $m(F_n) = \frac{1}{2} m(A_n)$, and if $n = 3, 5, 7, \cdots$, then $m(F_n) = \frac{1}{2} \left(1 - \frac{1}{n} \right) m(A_n)$. It follows that $m(F_n) \geqq \frac{1}{3} m(A_n)$ whenever $n \neq 1$. If F is the union of all the F_n's (i.e., $F = F_0 \cup F_2 \cup F_3 \cup \cdots$—note that there is no F_1), then F is disjoint from TF and $m(F) \geqq \frac{1}{3} (1 - m(A_1))$. Consequently

$$d(I, T) \geqq m(F + TF) \geqq \frac{2}{3} (1 - m(A_1)) = \frac{2}{3} d'(I, T) ;$$

the proof is complete.

The translations of the unit interval by $\frac{1}{3}$ and by $\frac{1}{2}$ (mod 1) show that both the bounds in the comparison theorem are best possible.

The comparison theorem tells us that the uniform topology is induced by either one of the metrics d or d'. The possibility of choosing whichever one is convenient at any given time frequently facilitates the proofs of properties of the uniform topology. Thus, for instance, a glance at d shows that the weak topology is smaller than the uniform topology. From this it follows that G is complete in the uniform topology. Indeed, if $\{T_n\}$ is a d-Cauchy sequence, then $\{T_n\}$ is weakly (ambidextrously) Cauchy, and therefore T_n tends to T, say, in the weak topology; it is easy to prove from here that T_n tends to T in the uniform topology. (The pointwise limit of a uniformly Cauchy sequence is necessarily a uniform limit.) On the other hand to prove that G is not separable in the uniform topology is most easily done by the use of d': if T_a denotes the translation of the unit interval by a (mod 1), then $d'(T_a, T_b) = 1$ whenever $a \neq b$. This proves, incidentally, that the uniform topology is strictly bigger than the weak topology.

UNIFORM APPROXIMATION

To get an approximation theorem for the uniform topology, it is convenient to lean on the following result, which has some independent interest.

LEMMA. *If E and F are Borel sets of the same measure in the interval X, then there exists an invertible measure-preserving transformation T on X such that $m(TE+F)=0$.*

PROOF. There are many ways to prove (and, incidentally, to improve) this lemma; some of them would take us on a Polish detour through the teratology of Borel sets. An easily available measure-theoretic proof runs as follows. If $m(E)=m(F)=0$, there is nothing to prove. Otherwise, let S be an arbitrary invertible ergodic transformation on X. For some positive integer n, the measure of the set $S^n E \cap F$ is not zero; define T on $E \cap S^{-n}F$ to be S^n. Now proceed inductively. If $m(E-S^{-n}F) \neq 0$, then apply the same reasoning to find a positive chunk of $E-S^{-n}F$ that some power of S maps into $F-S^n E$, and define T on that chunk to be that power of S. The transformation T is defined on E by a possibly transfinite repetition of this procedure, i.e., by the method of exhaustion. An application of the same process to $X-E$ and $X-F$ yields the extension of T to all X.

Observe that this lemma implies that the measure-theoretic structure of every Borel set is the same as that of an interval. It follows, in particular, that on every Borel set of positive measure there exists an ergodic measure-preserving transformation.

The analog of the weak approximation theorem for the uniform topology is the assertion that in the uniform topology the set of periodic transformations is everywhere dense in G. This assertion is an immediate consequence of the decomposition of a transformation into antiperiodic and periodic pieces and of the following quantitative assertion about antiperiodic transformations.

74

UNIFORM APPROXIMATION THEOREM. *If T is an antiperiodic transformation, then for every positive integer n and for every positive number ε there exists a transformation S such that S has everywhere the period n and such that* $d'(S, T) \leqq \frac{1}{n} + \varepsilon$.

PROOF. Apply Lemma 2 of the preceding section to obtain a measurable set E such that the sets $E, TE, \cdots, T^{n-1}E$ are pairwise disjoint and such that $m(E \cup TE \cup \cdots \cup T^{n-1}E) > 1 - \varepsilon$. If x is in $E \cup TE \cup \cdots \cup T^{n-2}E$, write $Sx = Tx$; if x is in $T^{n-1}E$, write $Sx = T^{-n+1}x$. The transformation S is defined thereby on $E \cup TE \cup \cdots \cup T^{n-1}E$. On its domain S is everywhere periodic of period n, and, no matter how S is extended to X,

$$d'(S, T) \leqq m(T^{n-1}E) + \varepsilon = m(E) + \varepsilon \leqq \frac{1}{n} + \varepsilon.$$

All that remains, therefore, is to define S on $X - (E \cup TE \cup \cdots \cup T^{n-1}E)$ so as to make it everywhere periodic of period n, and the possibility of doing so is an easy consequence of the lemma at the beginning of this section.

The first version of the uniform approximation theorem was a lemma in the proof of the category theorems to be discussed in the next section (Annals, 1944, p. 786); that version had $\frac{4}{n}$ in place of $\frac{1}{n} + \varepsilon$. The present version was stated (without proof) by Rohlin (Doklady, 1948, p. 349).

The theorem has an interesting corollary; I assert that in the uniform topology the set of all ergodic transformations is nowhere dense in G. Indeed, I know that every sphere in G contains a periodic transformation S, of period n, say. Let r be a positive number less than $1/n$ and such that the sphere K with center at S and radius r is entirely included in the given sphere; to prove the assertion of nowhere density, it is sufficient to prove that no transformation in K is ergodic. To prove this, suppose that $T \in K$ and write $E = \{x \colon Sx \neq Tx\}$. It follows that $m(E) < 1/n$; if $E^* = E \cup SE \cup \cdots \cup S^{n-1}E$, then $m(E^*) < 1$ and $SE^* = E^*$. If $x \in X - E^*$, then $Sx = Tx$; this implies that $T(X - E^*) = X - E^*$. Since $m(X - E^*) > 0$, the transformation T can be ergodic only if $m(E^*) = 0$. This, however, implies that $m(E) = 0$, and hence that $T = S$; the transformation T cannot be ergodic in any case.

A glance at the metric d' shows that it is very hard for two transformations to be near to each other in the uniform topology. In other words, the uniform topology is very large (i.e., has many open sets); it is so large, in fact, that it is virtually discrete. Consequently the assertion that some particular set (such as the set of periodic transformations) is everywhere dense in the uniform topology reveals a deep structural property of measure-preserving transformations. By the same token, however, it is very easy for any particular set (such as the set of ergodic transformations) to be nowhere dense; an assertion of topological smallness (e.g., being nowhere dense or being of the first category) is essentially a property of the topology only.

CATEGORY

I return now to the consideration of the weak topology. I want to say something about the topological size of three important sets of transformations, namely, the strongly mixing transformations, the weakly mixing transformations, and the ergodic transformations.

FIRST CATEGORY THEOREM. *In the weak topology the set of all strongly mixing transformations is a set of the first category.*

PROOF. Let P_k be the set of all periodic transformations of period k (i.e., the transformations T for which $T^k = I$), and write P_n^* for the union (over all $k > n$) of the sets P_k. The weak approximation theorem implies that each of the sets P_n^* is everywhere dense.

Let E be the first half of the unit interval and write $M_k = \{T: |m(E \cap T^k E) - 1/4| \leq 1/5\}$. The set M_k is closed in the weak topology. If M_n^* is the intersection (over all $k > n$) of the sets M_k, and if M^* is the union (over all n) of the sets M_n^*, then M^* contains every strongly mixing transformation; it is, therefore, sufficient to prove that M^* is of the first category. For this, in turn, it is sufficient to prove that each M_n^* is nowhere dense; since M_n^* is closed, this is equivalent to proving that $G - M_n^*$ is everywhere dense. One more reduction: since $G - M_n^*$ is the union (over all $k > n$) of the sets $G - M_k$, it is sufficient to prove that $P_k \subset G - M_k$. This, finally, is trivial: if $T \in P_k$, then $T^k E = E$, and therefore $m(E \cap T^k E) - 1/4 = 1/4$, so that T does not belong to M_k.

To study the class of weakly mixing transformations, I need a lemma that is interesting on its own right.

CONJUGACY LEMMA. *In the weak topology the conjugate class of each antiperiodic transformation T_0 (i.e., the set of all transformations of the form $S^{-1}T_0S$) is everywhere dense in G.*

PROOF. Let $N = \{T: m(PD_i + TD_i) < \varepsilon, i = 1, \cdots, n\}$ be a dyadic neighborhood of a permutation P; it is to be proved that for some S in G the transform $S^{-1}T_0S$ belongs to N. Write M for the weak neighborhood defined the same way as N except that $\varepsilon/2$ appears in the role of ε. The weak approximation theorem implies that M contains a cyclic permutation Q of rank k greater than the ranks of all D_i and such that $(1/2^{k-2})$ $<\varepsilon$. An application of the uniform approximation theorem (with 2^k and $1/2^k$ in place of n and ε, respectively) yields a transformation R such that R has (almost) everywhere the period 2^k and such that $d'(R, T_0)\leqq 2/2^k<\varepsilon/2$.

I assert that Q and R are conjugate in G. To prove this, denote by $E_0,\cdots, E_{q-1}(q=2^k)$ the dyadic intervals of rank k, arranged so that $QE_i=E_{i+1}$ (i mod q), and denote by T_0 a set of measure $1/q$ such that the sets F_0, RF_0, $\cdots, R^{q-1}F_0$ are pairwise disjoint. Write $F_i=R^iF_0, i=1,\cdots, q-1$, and let S be any measure-preserving transformation that carries E_0 onto F_0. The transformation S can be extended to the entire interval by writing $Sx = R^iSQ^{-i}x$ whenever x is in E_i. Schematically:

$$
\begin{array}{ccccccccc}
E_0 & \xrightarrow{Q} & E_1 & \xrightarrow{Q} & E_2 & \xrightarrow{Q} \cdots \longrightarrow & E_{q-2} & \xrightarrow{Q} & E_{q-1} \\
S\downarrow & & S\downarrow & & S\downarrow & & S\downarrow & & S\downarrow \\
F_0 & \xrightarrow[R]{} & F_1 & \xrightarrow[R]{} & F_2 & \xrightarrow[R]{} \cdots \xrightarrow[R]{} & F_{q-2} & \xrightarrow[R]{} & F_{q-1}
\end{array}
$$

It is easy to verify that $Q=S^{-1}RS$.

Now the proof is almost over. Since $d'(R, T_0)<\frac{\varepsilon}{2}$ and since d' is invariant under the group operations, it follows that

$$d'(Q, S^{-1}T_0S)=d'(S^{-1}RS, S^{-1}T_0S)<\frac{\varepsilon}{2}.$$

From this I conclude that

$$
\begin{aligned}
m(PD_i+S^{-1}T_0SD_i) &\leqq m(PD_i+QD_i)+m(QD_i+S^{-1}T_0SD_i) \\
&\leqq m(PD_i+QD_i)+d(Q, S^{-1}T_0S) \\
&<\frac{\varepsilon}{2}+d'(Q, S^{-1}T_0S)<\varepsilon.
\end{aligned}
$$

Consequently $S^{-1}T_0S\in N$; the proof is complete.

SECOND CATEGORY THEOREM. *In the weak topology the set of all weakly mixing transformations is an everywhere dense G_δ.*

PROOF. Let M denote the set of weakly mixing transformations. It is clear that every element of M is antiperiodic and that the set M is self-conjugate. Since, moreover, the set M is not empty, it is sufficient, in view of the conjugacy lemma, to prove that M is a G_δ.

For the purposes of this proof I shall denote the unitary operator induced by a measure-preserving transformation T by the symbol U_T. I shall need to make use of the fact that if f and g are in L_2, then the function whose value at T is $(U_T f, g)$ is continuous in the weak topology of G; this follows immediately from the fact that the weak topology of G is the restriction to G of the weak operator topology on the set of all unitary operators.

Let $\{f_n\}$ be a countable dense set in L_2; write

$$K(i, j, k, n) = \left\{ T : |(U_T^n f_i, f_j) - (f_i, 1)(1, f_j)| < \frac{1}{k} \right\}$$

and

$$K = \bigcap_i \bigcap_j \bigcap_k \bigcup_n K(i, j, k, n).$$

The fact that $(U_T f, g)$ depends continuously on T implies that each $K(i, j, k, n)$ is open, and hence that K is a G_δ. The characterization of weak mixing in terms of convergence outside a set of density zero shows that $M \subset K$; I shall complete the proof by showing that $K \subset M$. This is best done contrapositively; I shall show that if T is not mixing, then T does not belong to K.

If T is not mixing, then, by the mixing theorem, U_T has a non-trivial proper function. Without loss of generality I may assume that there exists a function f in L_2 and a constant c of absolute value 1 such that $\|f\| = 1$, such that $(1, f) = 0$, and such that $U_T f = cf$. Let i be such that $\|f - f_i\| < .1$; I shall show that T does not belong to K by showing that for this value of i, the transformation T does not belong to $K(i, i, 2, n)$ for any value of n. I must show, therefore, that

$$q_n = |(U_T^n f_i, f_i) - (f_i, 1)(1, f_i)| > .5$$

for all values of n.

Observe that $\|f_i\| \leq \|f\| + \|f - f_i\| < 1.1$. Since $|(U_T^n f, f) - (f, 1)(1, f)| = 1$, it follows that

$$1 \leq |(U_T^n f, f) - (U_T^n f, f_i)| + |(U_T^n f, f_i) - (U_T^n f_i, f_i)|$$
$$+ |(U_T^n f_i, f_i) - (f_i, 1)(1, f_i)|$$
$$+ |(f_i, 1)(1, f_i) - (f_i, 1)(1, f)|$$
$$+ |(f_i, 1)(1, f) - (f, 1)(1, f)|$$

$$\leq .1+.11+q_n+.11+0<q_n+.5;$$

the proof of the theorem is complete.

The second category theorem implies that in the weak topology the set of all weakly mixing transformations is of the second category; it follows, all the more, that the set of all ergodic transformations is of the second category. By a technique similar to the one used in the proof of the second category theorem it can also be proved that in the weak topology the set of all ergodic transformations is a dense G_δ; this is, if anything, easier than the corresponding fact for weakly mixing transformations.

Category theorems are often used in existence proofs. The second category theorem certainly does not belong to this category; the existence of a weakly mixing transformation was used in the course of the proof. The two category theorems together do, however, imply the existence of a transformation that is weakly mixing but not strongly mixing. The second category theorem also constitutes a solution of G.D. Birkhoff's conjecture that, in some sense, ergodic transformations represent the general case.

The second category theorem is historically prior to the first; I proved it in a paper entitled " In general a measure-preserving transformation is mixing " (Annals, 1944, p. 786). The first category theorem and its elegant proof were discovered by Rohlin (Dokıady, 1948, p. 349); the title of his paper is " In general a measure-preserving transformation is not mixing ".

The group G has received some further attention. For whatever amusement value it may possess, I report that Harada has proved that the group G is topologically simple (i.e., it has no non-trivial closed invariant subgroups) and arcwise connected (Japan Acad., 1951, p. 523).

INVARIANT MEASURES

Almost all the preceding work dealt with measure-preserving transformations; now I want to examine how likely it is that a transformation is measure-preserving. Suppose, as always, that X is a measure space with a sigma-finite measure m, and that T is a measurable transformation on X; for simplicity I shall assume that T is invertible. The problem is to find a measure p defined on the same class of measurable sets as m so that p is invariant under T.

The problem, as it stands, is not well formulated. What conditions, besides invariance, should the measure p satisfy? In a trivial sense the problem is always solvable; if p is identically zero, then p is an invariant measure. The only effective way of excluding this extreme case (and its equally annoying relatives) is to assume that p has no more sets of measure zero than m. In precise technical language, I shall require that the measure m be absolutely continuous with respect to p, i.e., that $m(E)$ should vanish whenever $p(E)$ vanishes. This delimiting of the problem is still not quite adequate. To illustrate the difficulty, let $p(E)$ be the number of points in E. It is clear that p is an invariant measure and that m is absolutely continuous with respect to p, but it is equally clear that p still constitutes a trivial solution of the problem. The way to cut out this p is to require that p be sigma-finite (just as m is sigma-finite). A still more drastic method is, of course, to require that p be finite.

The problem of invariant measure splits into two parts. Given an invertible measurable transformation T on a measure space X with sigma-finite measure m, when does there exist a finite measure p such that m is absolutely continuous with respect to p and such that p is invariant under T (Problem I), and when does there exist a sigma-finite measure p satisfying the same conditions (Problem II)? In order to solve these problems it is advisable to reduce them to some related and somewhat more special problems.

I assert, to begin with, that there is no loss of generality in assuming that m is finite. Suppose, indeed, that $\{E_n\}$ is a countable, disjoint class of sets of finite measure with union equal to X (such a class exists by the sigma-finiteness of m), and let $\{c_n\}$ be a corresponding countable class of positive numbers such that $\sum c_n m(E_n)$ converges. If $m_0(E) = \sum c_n m(E \cap E_n)$, then m_0 is a finite measure equivalent to m (i.e., m and m_0 have the same sets of measure zero). It follows that a necessary and sufficient condition that m be absolutely continuous with respect to a measure p is that m_0 be such. Replacing m by m_0, I may and do assume that m is finite ; as long as I am at it, I may as well assume that $m(X) = 1$.

The transformation T was assumed to be invertible but it was (obviously) not assumed to be measure-preserving. It could even happen that there exists a measurable set E of measure zero such that $m(T^{-1}E)$ is positive (or $m(TE)$ is positive, or both). A transformation for which this does happen is called singular ; if both $m(TE)$ and $m(T^{-1}E)$ vanish whenever $m(E)$ vanishes, the transformation T is called non-singular. I assert that from the point of view of the problem of invariant measure (either problem), there is no loss of generality in assuming that T is non-singular. To prove this, suppose that $\{c_n\}$ is a sequence of positive numbers, $n = 0$, $\pm 1, \pm 2, \cdots$, such that $\sum c_n = 1$; write $m_0(E) = \sum c_n m(T^n E)$. It is clear that m_0 is a normalized measure and that m is absolutely continuous with respect to m_0. I assert that if m is absolutely continuous with respect to some invariant measure p, then m_0 is also absolutely continuous with respect to p. Indeed, if $p(E) = 0$, then $p(T^n E) = 0$ for all n, and therefore $m(T^n E) = 0$ for all n ; this implies, as asserted, that $m_0(E) = 0$. From this remark it follows that the problem of finding p when m is given is the same as the problem of finding p when m_0 is given. If, however, $m_0(E) = 0$, then $m_0(TE) = m_0(T^{-1}E) = 0$; I may and do assume therefore that T is non-singular (with respect to m).

By two changes on m, I have ensured the finiteness of m and the non-singularity of T. My third and last comment of this type is that it is sufficient to restrict the search for an invariant p to the class of measures equivalent to m. I am saying, more precisely, that if p is a sigma-finite invariant measure such that m is absolutely continuous with respect to p, then there exists a sigma-finite invariant measure p_0 such that m is equivalent to p_0; if, moreover, p is finite, then p_0 will be finite also. To prove this, apply the Radon-Nikodym theorem to obtain a non-negative measurable function f such that $m(E) = \int_E f(x)dp(x)$ for every measurable set E. (It is true, but irrelevant, that since m is finite, the function f belongs to L_1 with respect to p.) If $A = \{x : f(x) = 0\}$, then $m(A) = 0$ and therefore, by non-singularity, $m(T^n A) = 0$ for all n ($n = 0, \pm 1, \pm 2, \cdots$). If B is the union (over all n) of the sets $T^n A$, then B is a measurable invariant set and $m(B) = 0$. If $p_0(E) = p(E - B)$, then p_0 is a measure; the measure p_0 is no less sigma-finite than p, and p_0 is invariant under T. If $p_0(E) = 0$, then $p(E - B) = 0$ and therefore $m(E - B) = 0$; since, however, $m(E - B) = m(E) - m(E \cap B) = m(E)$, it follows that $m(E) = 0$. If, conversely, $m(E) = 0$, then $m(E - B) = 0$. Since $E - B \subset X - B = \{x : f(x) > 0\}$, and since $m(E - B) = \int_{E - B} f(x)dp(x)$, it follows that $p(E - B) = 0$, and hence that $p_0(E) = 0$. This fulfills all my promises.

The problem (or problems) of invariant measure now reduce to this: given an invertible, measurable, and non-singular transformation T on a space of finite measure, find a finite (or else a sigma-finite) measure p equivalent to m and invariant under T.

INVARIANT MEASURES : THE SOLUTION

In a certain repulsive sense the problem of invariant measure is solved. In order to describe this sense, I must introduce a new concept. If $\{E_i\}$ is an infinite disjoint sequence of measurable sets with union E, if $\{F_i\}$ is another such sequence with union F, and if $\{n_i\}$ is a sequence of integers such that $T^{n_i}E_i = F_i$, then I shall say that the sets E and F are equivalent by countable decomposition. (It is clear that the relation thereby defined is indeed an equivalence.) If p is a measure invariant under T, then $p(E) = p(F)$ whenever E and F are equivalent.

In analogy with Dedekind's definition of finiteness, I shall say that a measurable set E is bounded if it is not equivalent by countable decomposition to a proper subset of itself; more precisely, E is bounded if the only measurable subsets of E that are equivalent to E by countable decomposition are (almost) equal to E. A set E is sigma-bounded if it is the union of a countable class of bounded sets. The transformation T is called bounded (or sigma-bounded) according as the whole space X is a bounded set (or a sigma-bounded set).

If the problem of invariant measure has a solution p, then every set of finite measure with respect to p is bounded; it follows that if p is finite, then T is bounded, and if p is sigma-finite, then T is sigma-bounded. The repulsive solution of the problem of invariant measure that I mentioned before is that the converses of these statements are also true: if T is bounded, then there exists a finite invariant measure equivalent to the given one, and if T is sigma-bounded, then there exists a sigma-finite measure with these properties. The finite assertion was proved by E. Hopf (Trans. A.M.S., 1932, p. 373), and the sigma-finite one by me (Annals, 1947, p. 735).

The concept of unboundedness is a kind of generalization of the concept of compressibility. More exactly, if E is compressible, then E is unbounded. Indeed, if $E \subset T^{-1}E$ and $m(T^{-1}E - E) > 0$, then $TE \subset E$

and $m(E - TE) > 0$ (recall that T was assumed to be non-singular); it follows that E is unbounded, and, all the more, that T is unbounded. From this, in turn, it follows that the problem of a finite invariant measure can have no solution if T is compressible. Suppose that T is incompressible —does the problem of finite invariant measure have a solution in that case? This question used to be considered non-trivial; it was an open question for fifteen years. The question is equivalent to this one : does there exist a conservative transformation T that preserves no equivalent finite measure? (The other assumptions are still in force : T is assumed to be measurable, invertible, and non-singular, as well as conservative.) By now the answer is known to be yes, and the proof is easy. Let T be an ergodic measure-preserving transformation on the line. (The line, to be sure, is not a space of finite measure, but this is inessential : Lebesgue measure can be replaced, if desired, by an equivalent finite measure. To simplify the discussion of the example, I do not make this replacement.) It is clear that T is measurable, invertible, and non-singular; I shall prove that it is conservative and that it preserves no equivalent finite measure.

If F is a measurable set disjoint from $T^{-n}F$ for $n = 1, 2, 3, \cdots$, then the sets $T^n F$ are pairwise disjoint $(n = 0, \pm 1, \pm 2, \cdots)$. If F has positive measure, then there exists a measurable subset G of F such that $0 < m(G) < m(F)$. The union of all the sets $T^n G$, $n = 0, \pm 1, \pm 2, \cdots$, is a non-trivial invariant set; since this contradicts the ergodicity of T, it follows that T is indeed conservative.

Suppose now that p is a finite invariant measure equivalent to m. It follows from the Radon-Nikodym theorem that there exists a non-negative function f integrable with respect to m such that $p(E) = \int_E f(x) \, dm(x)$ for every measurable set E. This implies that $p(TE) = \int_E f(Tx) \, dm(x)$; from this and from the invariance of p, I conclude that f is (almost) invariant under T. The ergodicity of T implies, therefore, that f is (almost) a constant; since f is integrable with respect to Lebesgue measure on the line, f must be equal to zero almost everywhere. It follows that p is identically zero, in contradiction to the assumption that p is equivalent to m.

The proof above did not use the concept of boundedness. The reason the "solution" of the problem of invariant measure in terms of that concept is unsatisfactory is that there does not appear to be any way of using the concept. Hopf's theorem on the subject was shown to be non-vacuous (i.e., it was shown that there exist transformations that are un-bounded in a non-trivial way) by using the concept of finite invariant measure itself. My theorem on the subject is still not known to be non-vacuous; for all anybody knows, every measurable, invertible, and non-singular transformation is sigma-bounded. I conjecture that this is false; it seems very probable that non-sigma-bounded transformations do exist. Since, however, I have no faith in the concept of boundedness as a tool in solving the problem, I shall not take the time to prove the two theorems involving that concept; I shall go on, instead, to reformulate the unsolved problem in two or three more hopeful ways.

INVARIANT MEASURES : THE PROBLEM

The first (and least interesting) reformulation of the problem of invariant measure is this : does there exist a completely unbounded transformation? (By a completely unbounded transformation I mean one that possesses no bounded sets of positive measure.) To prove that this question is equivalent to the problem of invariant measure, I argue as follows. A sigma-bounded transformation is certainly not completely unbounded ; if, consequently, it were true that every transformation is sigma-bounded, it would follow that there are no completely unbounded transformations. Suppose, conversely, that there exists a transformation that is not sigma-bounded ; I assert that in this case there exists an invariant set of positive measure such that the restriction of the transformation to that invariant set is completely unbounded. Indeed, in the contrary case every invariant set of positive measure includes a bounded set of positive measure ; an easy exhaustion argument, together with the fact that the given measure is sigma-finite, implies that the transformation is sigma-bounded after all.

A more promising reformulation of the problem of invariant measure reduces that problem to the solution of a functional equation. Suppose, as before, that T is non-singular ; it follows from the Radon-Nikodym theorem that there exists a positive measurable function w such that $m(TE) = \int_E w(x)\,dm(x)$ for every measurable set E. The function w might well be called the Jacobian of the transformation T; it plays the same role for T as the absolute value of the Jacobian plays for a differentiable transformation of Euclidean space. The transformation T is measure-preserving if and only if w is equal to the constant 1 almost everywhere.

If p is a sigma-finite invariant measure equivalent to m, then there exists a positive measurable function f such that $m(E) = \int_E f(x)\,dp(x)$. It follows that $m(TE) = \int_E f(Tx)\,dp(x)$, while at the same time, from the

87

Jacobian equation, $m(TE) = \int_E w(x)f(x)dp(x)$. I conclude that $f(Tx) = w(x)f(x)$ almost everywhere. Suppose, conversely, that this functional equation has a positive measurable solution, and write $p(E) = \int_E (1/f(x)) \, dm(x)$. Clearly p is a sigma-finite measure equivalent to m; since, moreover,

$$p(TE) = \int_E (1/f(Tx)) \, dm(Tx) = \int_E (1/f(Tx)) w(x) \, dm(x) = p(E),$$

it follows also that p is invariant.

It is of interest to note that a positive (but not necessarily measurable) solution of the functional equation always does exist. All I have to do is to define $f(x)$ arbitrarily for one point x out of each of the orbits of T, and use the functional equation to define f at the iterated images of x. If T is measure-preserving (in which case, to be sure, there is no point in considering the functional equation), then the equation becomes $f(Tx) = f(x)$, i.e., the equation of an invariant function, and any positive constant is an acceptable solution. In general, the product of an invariant function and a solution of the functional equation is another solution, and, moreover, every solution may be obtained from a particular one by multiplying it by a suitable invariant function.

The existence of completely unbounded transformations or solutions of the functional equation: these are two reformulations of the main problem. I turn now to a structure theorem that exhibits the way in which an arbitrary non-pathological transformation can be obtained from transformations for which the problem of invariant measure has a trivial solution.

From the point of view of searching for an invariant measure there are two particularly easy classes of transformations. One of these is the class of measure-preserving transformations—for these there is no problem. The other is the class of simple transformations, where a transformation T is called simple if there exists a measurable set E whose iterated images are all disjoint from one another and if the restriction of T to the complement of their union is the identity. (In other words, T is simple if its incompressible, or conservative, part is the identity.)

FACTORIZATION THEOREM. *Every measurable, invertible, and non-*

singular transformation T on the unit interval is a product, $T = PS$, where P is measure-preserving and S is simple.

PROOF. For every x in $[0, 1]$, write $Sx = m(T[0, x])$. It is clear that S is a monotone non-decreasing function such that $S0 = 0$ and $S1 = 1$. If $a \leq b$ and $Sa = Sb$, then $m(T(a, b]) = 0$. It follows that $m((a, b]) = 0$, so that $a = b$, and, consequently, S is strictly increasing.

If the numbers x_n converge to x_0 from above, then the sets $[0, x_n]$ form a decreasing sequence with intersection $[0, x_0]$; it follows that S is continuous on the right. If the numbers x_n converge to x_0 from below, then the sets $[0, x_n]$ form an increasing sequence with union $[0, x_0)$; it follows that S is continuous on the left. Consequently S is a homeomorphism of the interval onto itself; it follows automatically that S is an invertible and (Borel) measurable transformation.

Since $m(S[0, x]) = Sx = m(T[0, x])$, it follows easily that $m(SE) = m(TE)$ whenever E is an interval (with or without either of its endpoints), and hence that $m(SE) = m(TE)$ for every Borel set E. (From this it is easy to deduce that S is non-singular. It is also true that S is measurable in the Lebesgue sense as well as the Borel sense. I shall not need these facts.) If $P = TS^{-1}$, then P is an invertible, measurable transformation; since $m(PE) = m(TS^{-1}E) = m(SS^{-1}E) = m(E)$, so that P is measure-preserving, it remains only to prove that S is simple.

The unit interval is the union of the three sets $\{x : Sx = x\}$, $\{x : Sx < x\}$, and $\{x : Sx > x\}$, and each of the latter two, being open, is the union of countably many open intervals whose endpoints are in the first set. If $Sa < a$, then the iterated images under S of the interval $(Sa, a]$ are all disjoint from one another and exhaust one of the open intervals of the set $\{x : Sx < x\}$. For every open interval of each of the sets $\{x : Sx < x\}$ and $\{x : Sx > x\}$ such a generating subinterval can be found; it is clear that the union of these generating subintervals is a dissipative set whose iterated images cover all the dissipative part of the interval. This completes the proof of the factorization theorem.

What can be said about the Jacobian of the kind of transformation for which the problem of invariant measure is still open? The following result gives a certain amount of negative information.

JACOBIAN THEOREM. *If T is a measurable, invertible, non-singular,* *and incompressible transformation on a measure space X with a sigma-finite (but not necessarily finite) measure m, and if $m(TE) \leq m(E)$ for every measurable set E, then T is measure-preserving.*

PROOF. If T is not measure-preserving, then there exists a measurable set E such that $m(TE) < m(E) < \infty$. If $x \in E$, write $n(x)$ for the least positive integer such that $T^{n(x)}x \in E$; observe that, by the recurrence theorem, $n(x)$ is finite (almost) everywhere in E. Write $X_k = \{x \in E : n(x) = k\}$ and $F = \bigcup_{k=1}^{\infty} \bigcup_{i=0}^{k-1} T^i X_k$. It follows that $TF = \bigcup_{k=1}^{\infty} \bigcup_{i=1}^{k-1} T^i X_k \cup \bigcup_{k=1}^{\infty} T^k X_k \subset F \cup E = F$ and that $F - TF = E - \bigcup_{k=1}^{\infty} T^k X_k$. Since the assumption about E implies that $m(T^k X_k) < m(X_k)$ for at least one value of k, it follows that $m(F - TF) > m(E) - \sum_{k=1}^{\infty} m(X_k) = 0$, and hence that T is compressible. (The idea of this proof was shown to me by D.S. Ornstein.)

It follows from this theorem that the Jacobian of an interesting transformation cannot be identically a constant less than 1. In other words, a uniformly measure-contracting transformation is not interesting, and (consider inverses) neither is a uniformly measure-expanding transformation.

GENERALIZED ERGODIC THEOREMS

The first generalization I want to discuss is the so-called random ergodic theorem (see Pitt, Proc. Camb. Phil. Soc., 1942, p. 325, and Ulam and von Neumann, Bull. A.M.S., 1945, p. 660). The simplest example of such a theorem is obtained by considering two measure-preserving transformations S and T (on the same space X) and inquiring after the behavior of sequences such as $\left\{\frac{1}{n}\sum_{j=0}^{n-1}f(Q_j x)\right\}$, where each Q_j is obtained from its predecessor by multiplication either by S or by T. If, at each stage, the multiplier is " chosen at random ", then it becomes meaningful to ask for the probability that the sequence converge almost everywhere.

The correct way to construct a mathematical model for an infinite sequence of random choices (from two, or, for that matter, from any number of objects) is very well known. Let Y be a measure space with a probability (i.e., normalized) measure, and let Y^* be the space of all sequences (unilateral, say) of elements of Y, endowed with the infinite product measure (p, say). Suppose that to each point y in Y there corresponds an invertible measure-preserving transformation T_y on a measure space X with measure m. If $y^* = \{y_n\}$ is in Y^* ($n = 0, 1, 2, \cdots$), write $T_{y^*}^{(k)}$ for the product $T_{y_k}T_{y_{k-1}}\cdots T_{y_1}T_{y_0}$. Here $k = 1, 2, \cdots$; to define $T_{y^*}^{(k)}$ when $k = 0$, write $T_{y^*}^{(0)} = T_{y_0}$. If f is a complex-valued function on X, then, for each fixed y^*, it makes sense to ask about the almost everywhere convergence of $\frac{1}{n}\sum_{j=0}^{n-1}f(T_{y^*}^{(j)}x)$ and to ask about the measure of the set of y^*'s for which almost everywhere convergence does take place.

The formation of the sequence $\{T_{y^*}^{(k)}\}$ can be described differently. If S is the (unilateral) shift on Y^*, then $T_{y_k} = T_{y_0(S^k y^*)}$. If, in other words,

I write $T_{y\bullet}^{*} = T_{y_0}$, then $T_{y_k} = T_{Sky\bullet}$, and consequently

$$T_{y\bullet}^{(k)} = T_{S^k y\bullet}^{*} T_{S^{k-1}y\bullet}^{*} \cdots T_{y\bullet}^{*} T_{y\bullet}^{*}$$

for $k = 1, 2, \cdots$. (As before, $T_{y\bullet}^{(0)} = T_{y\bullet}^{*} = T_{y_0}$.) From this point of view the construction of Y^* is seen to be unnecessarily special; it serves only to define the measure-preserving transformation S. (I could have achieved the same result with an invertible transformation S simply by considering the bilateral sequence space.) If I forget about the source of Y^* (and if, to simplify the notation, I denote it henceforth by Y), I am led, finally, to the following precise generalization of the originally somewhat vaguely stated result. Let X and Y be measure spaces with normalized measures m and p, respectively. Suppose that S is a measure-preserving transformation on Y and that T_y is, for each y in Y, a measure-preserving transformation on X. Assume that the family $\{T_y\}$ is measurable, in the sense that the mapping (from $X \times Y$ into X) that sends (x, y) onto $T_y x$ is a measurable transformation. Write $T_y^{(0)} = T_y$ and $T_y^{(k)} = T_{S^k y} T_{S^{k-1}y} \cdots T_{Sy} T_y$ when $k \geq 1$.

RANDOM ERGODIC THEOREM. *If f is an integrable function on X, then, for almost every y, the averages* $\dfrac{1}{n} \sum_{j=0}^{n-1} f(T_y^{(j)} x)$ *converge almost everywhere; the limit function f_y^* is integrable.*

PROOF. I shall prove more than necessary; the "more" itself constitutes a possibly interesting generalization of the random ergodic theorem. Consider the transformation Q on $X \times Y$ defined by $Q(x, y) = (T_y x, Sy)$. The measurability hypothesis on $\{T_y\}$ implies that Q is measurable and measure-preserving. It is, in fact, the Cartesian product of two transformations: (1) the transformation (from $X \times Y$ into X) that sends (x, y) onto $T_y x$, and (2) the transformation (from $X \times Y$ into Y) that sends (x, y) onto Sy. It follows from the ergodic theorem that if g is an integrable function on $X \times Y$, then $\dfrac{1}{n} \sum_{j=0}^{n-1} g(Q_j(x, y))$ converges almost everywhere to a finite limit $g^*(x, y)$; the limit function g^* is integrable and $\iint g^*(x, y) \, dm(x) \, dp(y) = \iint g(x, y) \, dm(x) \, dp(y)$. This is the promised dividend; from this the stated version of the random ergodic theorem is a corollary.

Observe that (by an easy induction) $Q^k(x, y) = (T_y^{(k-1)}x, S^k y)$, $k = 1, 2, \cdots$. If g is defined by $g(x, y) = f(x)$, then $g(Q^k(x, y)) = f(T_y^{(k-1)}x)$. Everything is proved; as an extra dividend I get the fact that $\int f(x)\, dm(x) = \int\int f_y^*(x)\, dm(x)\, dp(y)$ (where f_y^* means what it obviously should).

A non-trivial question associated with this circle of ideas concerns the ergodicity and mixing properties of the auxiliary transformation Q. In certain special cases the ergodicity of Q can be characterized directly in terms of S and $\{T_y\}$; the general problem remains open. (See Kakutani, Berkeley Symposium, 1951, p. 247, and Gladysz, Bull. Polon., 1954, p. 411.)

The method of proof of the random ergodic theorem (i.e., the consideration of a skew product transformation such as Q) has other applications. Here is a sample: if S is a measure-preserving transformation on Y and if q is a measurable function of constant absolute value 1, then the modified averages $\dfrac{1}{n}\sum_{j=0}^{n-1}\Big(\prod_{0\leq i<j} q(S^i y)\Big) f(S^j y)$ converge almost everywhere to a finite (and, in fact, integrable) limit (provided that the given function f is integrable on Y). To prove this, let X be the circle (with normalized Lebesgue measure) and write $T_y x = q(y)x$ for each y in Y. The hypotheses of the random ergodic theorem are satisfied. If $g(x, y) = xf(y)$, then, by the ergodic theorem, $\dfrac{1}{n}\sum_{j=0}^{n-1} g(Q^j(x, y))$ converges almost everywhere to a finite limit $g^*(x, y)$; the limit function g^* is integrable on $X \times Y$. (The transformation Q is, of course, defined by $Q(x, y) = (T_y x, Sy) = (q(y)x, Sy)$.) Since

$$\frac{1}{n}\sum_{j=0}^{n-1} g(Q^j(x, y)) = \frac{1}{n}\sum_{j=0}^{n-1}\Big(\prod_{0\leq i<j} q(S^i y)\Big) \cdot x \cdot f(S^j y),$$

it follows that, for almost all x, convergence takes place for almost all y. To complete the proof, find any one well-behaved x and divide by it.

Still another result of the same type is this: if X, Y, S, and f are as above, then, for each c in X, the averages $\dfrac{1}{n}\sum\limits_{j=0}^{n-1}c^jf(S^jy)$ converge for almost all y in Y. To prove this, consider the transformation Q on $X \times Y$ defined by $Q(x, y) = (cx, Sy)$. If $g(x, y) = xf(y)$, then $g(Q^n(x, y)) = g(c^nx, S^ny) = c^nxf(S^ny)$—the rest of the proof is obvious. A related comment is that the averages $\dfrac{1}{n}\sum\limits_{j=0}^{n-1}x^jf(S^jy)$ converge to a finite limit for almost all pairs (x, y). For the proof consider the transformation Q on $X \times Y \times X$ defined by $Q(x, y, z) = (x, Sy, xz)$, and note that if $g(x, y, z) = zf(y)$, then $g(Q^n(x, y, z)) = x^nzf(S^ny)$. The strongest result of this type was announced by Wiener and Wintner (Amer. J., 1941, p. 794); they assert that there exists a set E of measure zero in Y such that if y does not belong to E, then $\dfrac{1}{n}\sum\limits_{j=0}^{n-1}x^jf(S^jy)$ converges for every x. I never could understand their proof.

A third direction of generalization was first obtained by Hurewicz (Annals, 1944, p. 195); it is an asymptotic result for transformations that are not necessarily measure-preserving. If (as is usual) we restrict attention to an invertible measurable transformation T, then there is hardly any loss of generality in insisting that T be non-singular and incompressible as well. The reduction is discussed in some detail in the preceding sections dealing with the problem of invariant measure. Even if that reduction is not considered fully convincing here, it is a fact that the most general known ergodic theorem follows easily from the medium general one that I propose to state, i.e., from the one for invertible, measurable, non-singular, and incompressible transformations.

Write, as once before, $dmT = wdm$ (i.e., $m(TE) = \int_E w(x)\,dm(x)$). It follows easily that $dmT^n = w_ndm$, where $w_n(x) = \prod\limits_{0 \leqq j < n} w(T^jx)$, $n = 1, 2, \cdots$. Theorem: if f is integrable, then the weighted averages

$$\sum_{j=0}^{n-1}f(T^jx)w_j(x) \Big/ \sum_{j=0}^{n-1}w_j(x)$$

converge almost everywhere to a finite limit $f^*(x)$; the limit function f^*

is integrable. Here the measure m may be infinite; if it is assumed that it is finite, then $\int f(x)dm(x) = \int f^*(x)dm(x)$. The proof of this result differs from the proof of the Birkhoff theorem in unexciting detail only. A condensed proof can be found in my note on the subject (Proc. N.A.S., 1946, p. 156). As long as the problem of invariant measure remains unsolved, this direction of generalization is a suspicious one; for all anyone knows, it does not apply to any cases not already included in Birkhoff's theorem.

The three generalizations mentioned above can be stirred together. There are some random ergodic theorems modified by a factor function q in the measure-preserving case, there are also some modified ergodic theorems in the non-measure-preserving case, and, I have no doubt, there must be some easily available modified random ergodic theorems in the non-measure-preserving case. I omit the gory details.

UNSOLVED PROBLEMS

1. Since physicists are anxious to replace phase means by time means, and since the ergodic theorem apparently justifies their doing so, Birkhoff's result was originally advertised as the cure for many of the ailments of statistical mechanics. It was soon observed, however, that the problem was merely shifted from one place to another : instead of having to decide whether phase means may be replaced by time means, the physicist now had to decide whether the given transformation is ergodic. For this reason, at least, it would be interesting to have available a few theorems that assert that under suitable conditions a measure-preserving transformation is ergodic ; no general results of this type are known.

2. The recurrence theorem and the ergodic theorem indicate that if T is a measure-preserving transformation and if f is an integrable function, then almost every one of the sequences $\{f(T^n x)\}$ is " recurrent ", in some unspecified sense. Can one define a concept of recurrence for sequences of real numbers so that if $\{a_n\}$ is, in that sense, a recurrent sequence, then $\frac{1}{n}\sum_{j=0}^{n-1} a_j$ converges to a finite limit, and so that if T is measure-preserving and f is integrable, then $\{f(T^n x)\}$ is recurrent almost everywhere?

3. The outstanding algebraic problem of ergodic theory is the problem of conjugacy : when are two transformations conjugate? This is vague, of course, but there are some interesting and quite specific yes-or-no questions connected with it that should be solved. (The general problem might never be solved ; it is not even quite clear what it means.) Here are a few samples : (a) Do there exist measure-theoretically non-isomorphic spaces such that the shifts based on them are conjugate? (b) Do there exist two equivalent but non-conjugate transformations with continuous spectrum? (c) Do there exist two conjugate ergodic automorphisms of a compact group that do not belong to the same conjugate class in the group of all automorphisms of that group?

4. The problem of square roots is essentially still open, and so, of course, is the more general problem of n-th roots, and the problem of embeddability into a flow. When does a measure-preserving transformation have a square root? More specifically: does every transformation with a continuous spectrum have a square root? Does every shift have a square root?

5. What unitary operators are induced by measure-preserving transformations? Suppose, to look at a concrete special case, that the vectors g and $f_{i,j}$ constitute a complete orthonormal set in a Hilbert space, $i = 1$, $\cdots, n$, $j = 0, \pm 1, \pm 2, \cdots$, and that U is the unitary operator defined by $Ug = g$ and $Uf_{i,j} = f_{i,j+1}$. Does there exist a measure-preserving transformation T on a space of finite (!) measure such that the unitary operator induced by T is equivalent to U?

6. Liouville's theorem asserts the existence of an invariant measure; it should be embedded into a suitable general context. The problem is, of course, the problem of invariant measure; its apparent impregnability is humiliating. The simplest yes-or-no question was formulated above thus: if T is a measurable, invertible, and non-singular transformation on a measure space with measure m, does there exist a sigma-finite measure equivalent to m and invariant under T? The question can be formulated in group-theoretic language: a single transformation can be viewed as the generator of a cyclic group. What can be said about more general abelian groups? What about not necessarily abelian groups? Except for some trivial remarks that can be made about compact abelian groups, all that is known (von Neumann, Annals, 1940, p. 94) is that the answer to the question of the existence of an equivalent invariant measure is (for general groups) sometimes no. When it is yes?

7. Functional equations such as $f(Tx) = w(x)f(x)$ occur in the theory of invariant measures, and they occur also, in quite a different manner, in the theory of generalized proper values. (The function w is given, the function f is unknown.) A systematic approach to their study is badly needed. Sample yes-or-no question: does there exist a measure-preserving transformation T on a non-atomic space of finite measure such that the functional equation has a measurable solution f (of constant absolute value 1) for every measurable function w (of constant absolute value 1)?

8. In the hope of cracking the conjugacy problem, several conjugacy invariants for transformations have been proposed from time to time. An obvious one is the class of invariant subalgebras of the measure algebra. If B is the measure algebra associated with a given measure space X and if T is the automorphism of B induced by a measure-preserving transformation on X, a subalgebra B_0 of B is said to be invariant under T if $TE \in B_0$ whenever $E \in B_0$. ("Subalgebra" means "sub-sigma-algebra" here.) What are the possibilities for non-trivial invariant subalgebras? Specifically: suppose that X is the bilateral sequence space based on the two-point space $\{-1, +1\}$ and that T is the automorphism induced by the bilateral shift. Let B_0 be the class of all "symmetric" measurable sets (or, rather, the class of all equivalence classes of such sets modulo sets of measure zero), where a set E is called symmetric if and only if $\{-x_n\} \in E$ whenever $\{x_n\} \in E$. The subalgebra B_0 is invariant under T; does T have any other non-trivial invariant subalgebras?

9. A curious conjugacy invariant is suggested by automorphisms on compact abelian groups. Suppose that X is the torus and that T is the automorphism corresponding to the unimodular matrix $M = \begin{pmatrix} a & b \\ c & d \end{pmatrix}$. Write k for the trace of M, i.e., $k = a + d$, and, to simplify matters, assume that the determinant of M is $+1$. The fact that M (and also the transpose of M) satisfies the Hamilton-Cayley equation $(M^2 - kM + 1 = 0)$ implies that if f is a character of X, then $(U^2 f) \cdot (Uf)^{-k} \cdot f = 1$. (Here U is, of course, the unitary operator induced by T.) The existence or non-existence of f's satisfying such identities is a conjugacy invariant of T. Sample question: if N is another unimodular matrix, with corresponding automorphism S, such that the characteristic equation of N is different from that of M, and if V is the unitary operator induced by S, does there exist any non-constant function f of constant absolute value 1 such that $(V^2 f) \cdot (Vf)^{-k} \cdot f = 1$?

10. Rohlin (Izvestya, 1949, p. 329) has proposed the concept of r-fold mixing as a conjugacy invariant. Let r be a positive integer, $r = 1$, $2, 3, \cdots$, and consider ordered $(r+1)$-tuples K of non-negative integers, $K = (k_0, \cdots, k_r)$. (There might be coincidences among the k's.) Define the stretch of K, in symbols $s(K)$, as the minimum of all the differences $|k_i - k_j|$, where $i, j = 0, \cdots, r$, and, of course, $i \neq j$. An infinite sequence $\{K(n)\}$ of such $(r+1)$-tuples, $K(n) = (k_0(n), \cdots, k_r(n))$, is admissible if $s(K(n))$ tends to infinity with n. The transformation T is r-fold mixing if for every admissible sequence $\{K(n)\}$, and for every ordered $(r+1)$-tuple of measurable sets $(E_0, \cdots, E_r)$, it is true that $m\left(\bigcap_{i=0}^{r} T^{k_i(n)} E_i \right)$ converges, as n tends to infinity, to $\prod_{i=0}^{r} m(E_i)$. This invariant is not yet known to distinguish between any two transformations: Rohlin asserts that every ergodic automorphism of a compact group is r-fold mixing for all r. Question: do there exist mixing (i.e., 1-fold mixing) transformations that are not 2-fold mixing?

REFERENCES

Halmos, Measure theory, 1950.

Halmos, Introduction to Hilbert space, 1951.

E. Hopf, Ergodentheorie, 1937.

Khintchine, Mathematical foundations of statistical mechanics, 1949.

Pontrjagin, Topological groups, 1939.

In my A.M.S. address (Bull. A.M.S., 1949, p. 1015) there is a historical discussion of most of the subjects that I will take up in these lectures and there is a bibliography that is complete through 1948. References to other subjects that will be mentioned from time to time, but whose detailed treatment will not be taken up, can be found in the same address, or in Kakutani's 1950 report (Cambridge Congress, vol. 2, p. 128), or in Oxtoby's A.M.S. address (Bull. A.M.S., 1952, p. 116). Further references will be given at appropriate points during the lectures.